Pathology for Toxicologists

Pathology for Toxicologists

Principles and Practices of Laboratory Animal Pathology
for Study Personnel

Edited by Elizabeth McInnes

This edition first published 2017 © 2017 John Wiley & Sons Ltd

Registered office
John Wiley & Sons Ltd, The Atrium, Southern Gate, Chichester, West Sussex, PO19 8SQ, United Kingdom

For details of our global editorial offices, for customer services and for information about how to apply for permission to reuse the copyright material in this book please see our website at www.wiley.com.

Library of Congress Cataloging-in-Publication data is applied for

ISBN: 9781118755419 (hardback)
ISBN: 9781118755402 (paperback)

A catalogue record for this book is available from the British Library.

Cover image: Courtesy of the author
Cover design: Wiley

Set in 10/12pt Warnock by SPi Global, Chennai, India
Printed and bound in Malaysia by Vivar Printing Sdn Bhd
10 9 8 7 6 5 4 3 2 1

Contents

List of Contributors

Elizabeth McInnes
Cerberus Sciences
Thebarton, SA
Australia

Natasha Neef
Vertex Pharmaceuticals
Boston, MA
USA

Cheryl L. Scudamore
MRC Harwell, Harwell Science and
Innovation Campus
Oxfordshire
UK

Bhanu Singh
Discovery Sciences
Janssen Research & Development
Spring House, PA
USA

Barbara von Beust
Independent consultant
Winterthur
Switzerland

Preface

Seemingly minor differences in opinion between the study pathologist and the study director or vice president of safety assessment (positions often filled by toxicologists) can escalate to cause real problems with the development of a test article in the pharmaceutical and agrochemical industries. Pathology is an imprecise science that relies on the observation of subtle variations in patterns of cellular arrangement and the tinctorial affinity of certain cells for staining procedures. Non-pathologists, such as toxicologists, can find it difficult to understand the data they receive from pathologists, partly due to the subjectivity of the discipline and the variations between pathologists, and partly due to the terminology that pathologists use.

There is a lack of pathological texts aimed at study personnel. This book has been written for toxicologists at all stages of their training or career who want to know more about the pathology encountered in laboratory animals, including study directors, study monitors, undergraduate and postgraduate toxicology students, toxicology report reviewers and research scientists employed in the pharmaceutical industry. The aim is to help study personnel bridge the gap in the understanding of pathology data. The book will enable them to understand the pathology reports they receive and the common pathologies encountered, so that they can more easily integrate pathology data into their final study report and ask pathologists relevant questions where there are gaps in understanding.

We have attempted to make the book user-friendly and easy to understand. Important lesions in rats, mice, non-human primates, mini pigs and beagle dogs, the most common laboratory animals used in the industry, are discussed. The compound-induced pathology in all the major organ systems is covered, as are clinical pathology, adversity and the limitations of pathology. There is also a glossary, which should help all non-pathologists understand the language of toxicological pathology. The aim is to demystify such terms as "chronic focal hepatic hypertrophy with Ito cell tumor."

This book is intended to give study personnel an insight into the uncertainties encountered by the pathologist when reading studies and to provide them with explanations for why pathologists cannot always make up their minds. We trust it will improve communication and understanding between pathologists, toxicologists and study directors, so that a more succinct and helpful toxicologist report can be written.

Elizabeth McInnes

1

An Introduction to Pathology Techniques

Elizabeth McInnes

Cerberus Sciences, Thebarton, SA, Australia

Learning Objectives

- Understand the role of study personnel in necropsies.
- Understand the various steps involved in producing glass slides from harvested tissues.
- Understand the ancillary techniques used in toxicological pathology.
- Discover what carcinogenicity, inhalation and crossreactivity studies entail.

This book is aimed at all study personnel – including study monitors, study directors and toxicologists – who are exposed to pathology reports, necropsies, peer review, haematology and biochemistry results and adversity on a regular basis. The secret to an informative, relevant and useful pathology report is an open and collegial relationship between the study director and the study pathologist (Keane, 2014). This chapter aims to describe the various stages of the pathological process (e.g. necropsy, fixation of tissues, cutting of slides) in order to demonstrate where crucial errors which can cause problems at a later stage, may arise. In addition, it includes a brief overview of ancillary techniques that pathologists sometimes use (e.g. electron microscopy). Finally, it discusses carcinogenicity studies, digital pathology, biological drugs and crossreactivity studies, and their impact on study personnel. Throughout the chapter, the client is referred to as the 'sponsor' of a particular pharmaceutical study.

Pathology is the study of disease, particularly the structural and functional changes in tissues and organs. Toxicological pathology is concerned predominantly with cell and tissue injury in animals treated with introduced chemical compounds or biological drugs. Studies are regulated by international bodies such as the Organisation for Economic Co-operation and Development (OECD), the US Food and Drug Administration (FDA) and the European Medicines Agency (EMA). Animal testing to determine the safety of pharmaceuticals, medical devices and food/colour additives is required by the FDA before it will give approval to begin clinical trials in humans. Pathology data may be quantitative (haematology, chemistry data, organ weights) or qualitative (microscopic diagnoses), and the toxicological pathology report is divided into macroscopic and

Pathology for Toxicologists: Principles and Practices of Laboratory Animal Pathology for Study Personnel,
First Edition. Edited by Elizabeth McInnes.

microscopic findings. Study personnel are ultimately responsible for the study report, including the pathology report and data. Thus, study personnel need to understand what the pathology report means and how it has been generated. This chapter aims to help the study director understand the processes involved in a study, from harvesting tissues from the animal to generating glass slides to producing a pathology report.

1.1 Animal Considerations

The main species of animal used in the pharmaceutical industry are rats, mice, dogs, non-human primates, minipigs and rabbits. Occasionally, farm animals, hamsters, cats and gerbils are also used. There are no absolute reasons for selecting a particular animal species for systemic toxicity, but for acute oral, intravenous, dermal and inhalation studies and studies of medical devices, the mouse or rat is preferred, with the option of the rabbit in the case of dermal and implantation studies. Non-rodent species may also need to be considered for testing, although a number of factors might dictate the number and choice of species. Carcinogenicity studies generally use rats and mice.

All animal studies must be conducted according to the animal welfare laws of the country in which they are based, and in general studies may use protected animals only if there are no other reasonable, practicable choices for achieving a satisfactory result. Laboratory animals may only be used in minimal numbers, where they have the lowest possible degree of neurophysiological sensitivity and where the study causes minimal pain, suffering, distress and lasting harm. Animal suffering must be balanced against the likely benefits for humanity, other animals and the environment. In general, in the planning of all preclinical studies, due consideration must be given to reduction, refinement and replacement (the '3 Rs') (Tannenbaum and Bennett, 2015).

Generally, only healthy purpose-bred young adult animals of known origin and with defined microbiological health status should be used in pathology studies (i.e. health monitoring of sentinel animals must be performed before the study starts). Health monitoring involves testing for the various bacteria, viruses, parasites and protozoa that may infect experimental animals and compromise study results (McInnes et al., 2011). The weight variation within a sex should not exceed 20% of the mean weight, and when female animals are used, they should be non-pregnant and should never have borne young.

Laboratory animals should be given a short acclimatisation period at the start of the study. Control of environmental conditions and diet and proper animal care techniques are necessary throughout the study in order to produce meaningful results. The number of animals needed per dose group depends on the purpose of the study. Group sizes should increase with the duration of treatment, such that at the end of the study sufficient animals are available in every group for a thorough biological evaluation and statistical analysis.

1.2 Necropsy

Necropsies or post mortem examinations (Figure 1.1) on experimental animals are a fundamental part of toxicological pathology (Fiette and Slaoui, 2011). They generally take place at the end of a study, but are also conducted if an animal dies early. The necropsy and

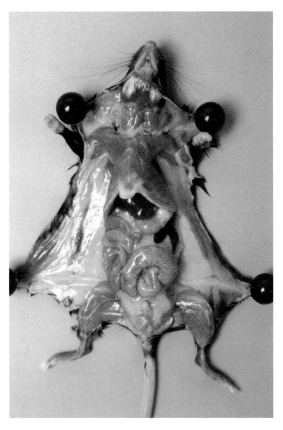

Figure 1.1 Necropsies or post mortem examinations on experimental animals such as this mouse are a fundamental part of toxicological pathology.

pathology data are the single most important aspect of the pathology process, and study personnel get only one chance to retrieve them: once the tissues have been discarded, potentially valuable information is lost forever. At the necropsy, all macroscopic findings and abnormalities visible to the naked eye (e.g. enlarged liver, ulcerated skin, the presence of diarrhoea) are noted and recorded. In addition, tissues are collected for examination under the microscope. The pathologist, necropsy supervisor, prosector, phlebotomist and weighing assistant are all responsible for the recording the macroscopic data (Keane, 2014). Some studies collect all tissues (a full tissue list is indicated in the study plan or protocol), while others harvest only a limited list. A copy of the study plan should be available in the necropsy room to ensure that study personnel collect the correct tissues. Sometimes, all the tissues are collected into formalin, but slides are only cut if the sponsor decides there is a need later on. Harderian glands and draining lymph nodes are examples of tissues that are not always collected: study personnel should check the study plan before beginning the necropsy.

Study directors and toxicologists are often required to attend the necropsies of the animals on their studies. Although not directly involved in the necropsy process, it is nevertheless important that study personnel understand the process and are able to provide advice and management, particularly if unusual tissues are to be collected, severe test article-related findings are observed in high-dose animals, or deviations from standard operating procedures (SOP) occur. The study pathologist may be consulted if there is an unusually high rate of unscheduled deaths in the study or it is difficult to characterise macroscopic findings (such as very white teeth noted when treatment causes defects in enamel formation) (Keane, 2014).

Carbon dioxide asphyxiation provides a rapid form of euthanasia for mice (Seymour et al., 2004) and rats, but it can cause severe lung haemorrhage, which may make microscopic examination of the lungs difficult. Barbiturate overdose is an effective form of euthanasia, but it requires the use of pentobarbital sodium (Seymour et al., 2004). Larger animals (e.g. rabbits, non-human primates and dogs) are euthanised by an overdose of sodium pentobarbitone, which may cause congestion of some organs and is highly irritant if injected into the tissues around the vein.

Control and treated animals should always be necropsied by the same team of technicians, and the animal numbers and order of examination should be randomised. The identity of an animal is given in a tattoo, ear notch or microchip and should be recorded on all necropsy storage containers in indelible ink.

Organs should be weighed at necropsy; increases and decreases in organ weight can often be correlated with the microscopic findings observed by the pathologist. To ensure meaningful organ weights are recorded, the organs should be taken from an exsanguinated animal (if possible), and excess moisture and adipose tissue should be removed. Guidelines on the weighing of organs are outlined in various papers (Michael et al., 2007; Sellers et al., 2007).

The macroscopic lesions observed at necropsy may be the only pathological data generated from certain studies and must be presented in the form of an incidence table. Consequently, lesions should be described consistently throughout the necropsy process, and standardised terms should be used. The use of an agreed, standardised macroscopic glossary will help to reduce the incidence of different personnel using different terms to describe the same lesion (Scudamore, 2014). In studies in which histopathology will be performed, the macroscopic lesions observed at necropsy are very important to the pathologist, as they often correlate with the lesions observed under the light microscope (e.g. an enlarged yellow liver at necropsy will often have lipid vacuolation in haepatocytes revealed under light microscopy).

Macroscopic lesions at necropsy should simply be described: no attempt at interpretation or diagnosis should be made at this stage (e.g. necropsy staff should not describe an enlarged, mottled liver as 'hepatitis' or a yellow tissue colour should not be described as jaundice). This is because once signed, the anatomic pathology report cannot be easily reinterpreted. All macroscopic lesions observed at necropsy should be described in terms of size and distribution (focal, multifocal and diffuse), colour and consistency (soft, friable, firm, hard, fluid filled, gritty, etc.). The location, size and number of a mass or lesion should be recorded. A standard diagram of the dorsal and ventral aspects of the animal is useful for recording the exact locations of lesions and masses. All measurements should be made in millimetres, and terms such as 'enlarged', 'pale' and 'small' should be avoided or should be accompanied by an actual measurement or colour. In particular studies, it may be useful to photograph certain lesions in order to illustrate their exact nature and severity to future study personnel. However, although photographs are a good record of macroscopic lesions observed at necropsy, there may be Good Laboratory Practice (GLP) and legal issues to contend with (Suvarna and Ansary, 2001).

Autolysis occurs within 10 minutes of the death of an animal (Pearson and Logan, 1978), so necropsy should be performed as quickly and efficiently as possible, with

Figure 1.2 Post mortem imbibition of bile pigment in the mesenteric fat tissue adjacent to the gall bladder. Photograph taken from a bovine necropsy.

limited tissue handling, squeezing and tissue damage. Post mortem change occurs as a result of autolysis (action of enzymes from the ruptured cells on the dead animal's cells) and putrefaction (degradation of tissue by the invasion of certain microorganisms); changes include *rigor mortis* (stiffening of limbs and carcase), clotting of the blood, hypostatic congestion (pooling of blood into the dependent side of the carcase, termed '*livor mortis*'), imbibition of blood (or bile pigment; Figure 1.2) and gaseous distension of the alimentary tract. In addition, pseudomelanosis (the greenish or blackish discolouration of tissues due to ferrous sulphide) tends to occur in organs that lie adjacent to the intestines, such as the liver. Most of these changes will be visible if an animal dies during the night or on the weekend, and every effort should be made to store the carcase in a fridge and to perform a necropsy as soon as possible thereafter.

1.3 Lung Inflation with Fixative

Tracheal instillation of the lungs with fixative at necropsy is required to improve the histology of the pulmonary architecture in mice and rats, and is recommended for all rodent studies. Tracheal instillation of fixative may be performed either after removing the lungs from the thoracic cavity or with the lungs in situ (Braber et al., 2010). It may sometimes be easier to inflate only one lung lobe, using a needle and syringe to inject formalin (Knoblaugh et al., 2011).

1.4 Fixation

In general, fixation of tissues maintains cellular integrity and slows the breakdown of tissues by autolysis. The most common fixative is 10% neutral buffered formalin, which ensures rapid tissue penetration, is easy to use and is inexpensive. However, formalin is

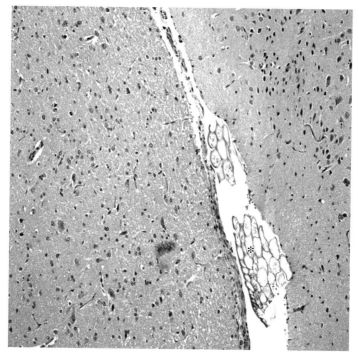

Figure 1.3 Artefacts induced at necropsy include inclusions of foreign material into the issue, such as plant material (*) during brain removal.

highly toxic and carcinogenic and may have effects on the immune system (Costa et al., 2013). Tissues should be fixed at a 1 : 10 or 20 ratio of fixative to tissue for at least 48 hours. Modified Davidson's is the recommended fixative for eyes and testes, as it prevents retinal detachment in the eye and separation of cells lining the seminiferous tubules in the testes. Glutaraldehyde or osmium tetraoxide is used for the fixation of tissues intended for electron microscopy. Artefacts which occur at necropsy include inclusions of foreign material into the tissue (e.g. plant material during brain removal (Figure 1.3) and the incorporation of sharp shafts of hair into soft tissues) and pressure and pinch effects (from forceps) (McInnes, 2011). These can be confused with lesions by an inexperienced pathologist.

1.5   Making Glass Slides

The production of glass slides suitable for histopathological analysis involves a number of steps performed in the histology laboratory (Figure 1.4).

1.5.1   Trimming

In the first step, formalin-fixed tissues collected during the necropsy are further cut up in order to fit into the embedding cassettes (Knoblaugh et al., 2011). Two steps in the pathology phase of the study cannot be repeated: the necropsy and the macroscopic tissue trimming. This is because if tissues are discarded after the necropsy or at trimming,

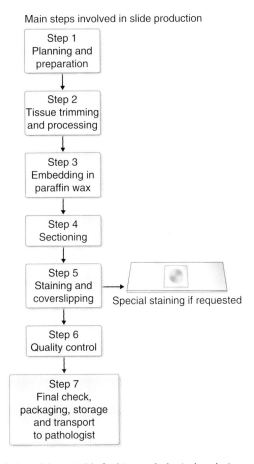

Figure 1.4 Production of glass slides suitable for histopathological analysis.

it is impossible to return to them. Great care should thus be exercised at the trimming stage. All tissues should be trimmed in the same way, from the same area in the organ, and all described gross lesions must be identified and included in the cassette (Figure 1.5). The cassette lid will cause impression marks on the tissue surface if the tissue is too big for the cassette (Knoblaugh et al., 2011). Excellent trimming and blocking patterns indicating how to section each tissue and which tissues should be placed together in a cassette are contained in Ruehl-Fehlert et al. (2003), Kittel et al. (2004) and Morawietz et al. (2004). The staff involved in trimming should be aware that variations in the incidence of certain lesions (such as thyroid C-cell findings and thyroid tumours) can be associated with the type of section taken (i.e. transverse compared to longitudinal).

It is essential that the cassette be marked correctly with the animal's identification number, sex and group, and with either the tissue name(s) or a number indicating which tissues are always trimmed into that particular cassette (Figure 1.6). The blocking sheet (Figure 1.6) indicates which tissue has been processed in which cassette. Multiple tissues may be grouped in one cassette (e.g. different tissues from the gastrointestinal tract), but certain tissues (e.g. adrenal and bone) should not be grouped together, since differences in consistency will cause problems during the microtoming of the wax

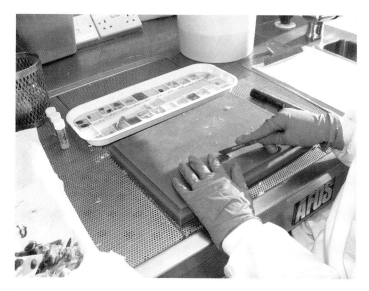

Figure 1.5 Tissue trimming and placement in a cassette.

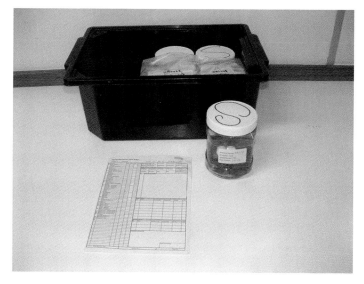

Figure 1.6 The blocking sheet.

blocks. The collection and histological examination of heart valves was overlooked in the past, but with the introduction of reporting of valvular lesions associated with the use of antiobesity drugs (fenfuramine-fentermine) (Connolly et al., 1997), these tissues must now be examined. Thus, histology technicians now endeavour to cut the heart into sections, which allows for the visualisation of the major aortic and pulmonic heart valves. Artefacts introduced at this stage include small pieces of tissue from the previous cassette superimposed on another tissue (due to failure to clean the knife between tissues and between animals) (Figure 1.7) (McInnes, 2011).

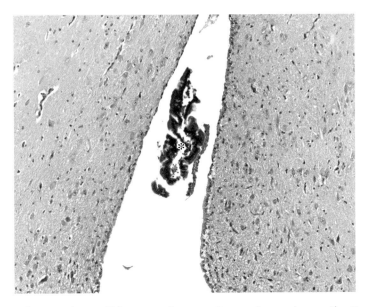

Figure 1.7 Small pieces of tissue (*) from a previous cassette superimposed on another tissue.

1.5.2 Tissue Processing

After trimming, the cassettes are placed in a machine that allows the tissues to undergo a series of steps which include tissue dehydration, clearing and impregnation with paraffin wax (Figure 1.8). Paraffin wax serves to keep tissue firm and intact, and in the correct orientation for sectioning.

1.5.3 Embedding

During embedding, a trained technician places the paraffin wax-infiltrated tissues and additional wax into a mould, which is chilled to produce tissue blocks.

1.5.4 Microtoming

During microtoming, thin sections (~4–6 µm) are cut from the wax block using a rotary microtome (Figure 1.9). The operation of this device is based upon the rotary action of a hand wheel, which advances the specimen (wax block) towards a rigidly held blade. The thin wax sections are then floated in a water bath, and appropriately identified glass slides are used to scoop them out of the water. The glass slides are placed in an oven to melt off the wax, leaving only the unstained tissue. Histology technicians require fine motor skills in order to ensure that all of the anatomic features of small tissues such as the adrenals and pituitary (e.g. cortex and medulla of adrenal) are displayed on a single slide. The formalin fixed wet tissues may be discarded after the study is finalised, however the wax tissue blocks and the glass slides are archived according to the protocol (Keane, 2014).

1.5.5 Staining

Before staining, all tissues are transparent, and it is hard to make out any cellular detail under the microscope. For this reason, stains are bound to different parts of the tissue,

Figure 1.8 Machine for tissue dehydration, clearing and impregnation with paraffin wax.

allowing these components to be visualised. Histochemical stains are made up of chemical dyes that bind to tissues by the same mechanism as a chemical interaction. The most commonly used stain is haematoxylin and eosin (H&E), which stains acidic tissue components pink (the eosin binds to the cytoplasm, making it pink) and basic tissues blue (the haematoxylin binds to the nuclei of the cells, making them blue) (Figure 1.10).

After staining, the section is mounted with a coverslip to prevent the tissue from drying out, prevent surface damage to the tissue and improve tissue transparency. If the pathologist discovers that a tissue is missing, all attempts must be made to find the missing tissue (by going back to the wet tissues, or by cutting deeper into the original block if no wet tissue remains); if a tissue is inadequate, it must be improved (by re-embedding). Occasional missing or inadequate tissues are acceptable in a study, but large numbers will compromise its completeness, and the study may have to be repeated – at great cost.

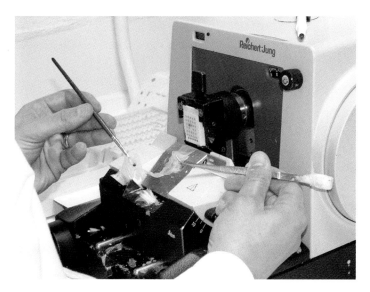

Figure 1.9 Rotary microtome.

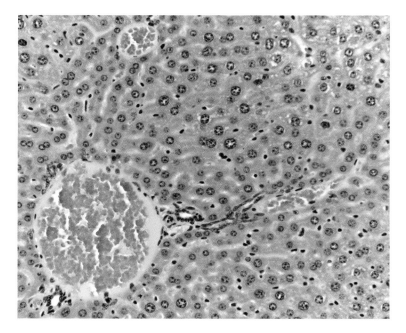

Figure 1.10 H&E staining.

1.5.6 Quality Control

The final step in slide processing is to carefully check the glass slide for artefacts, to make sure that the information on the block is the same as that on the slide and to ensure that the slides have been arranged in the correct order according to the blocking sheet (which pertains to one animal) (Figure 1.11).

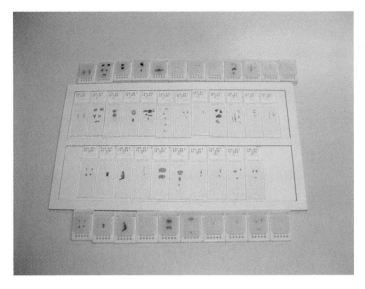

Figure 1.11 Quality control.

1.6 Special Histochemical Stains

Stains with specific affinities for different tissue components (e.g. calcium, fat) can be used to confirm the identity of these tissues or substances (see Table 1.1). Examples include phosphotungstic acid haematoxylin (PTAH) (Figure 1.12), which demonstrates muscle striations; periodic acid–Schiff (PAS), which stains glycogen and carbohydrate; Congo red, which stains amyloid; and Oil Red O, which stains lipid (can only be used on fresh-frozen tissue that has not been fixed in formalin). Pathologists may ask for special

Table 1.1 Special histochemical stains.

Substance	Tissue seen in	Special stain used
Bile	Liver	Fouchet's
Lipofuscin	Various (generally older animals)	PAS, Sudan black B, Schmorl's, Long Ziehl-Neelsen technique
Glycogen	Liver, muscle	PAS with diastase
Haemosiderin (golden brown)	Spleen, others	Perl's Prussian blue
Formalin pigment (artefact)	Blood-rich tissues, large areas of haemorrhage	- (need to extract it from sections using picric acid)
Melanin (more dark brown–black)	Lung, muscle, others	By exclusion
Fat	Liver	Oil Red O on frozen non-formalin fixed tissue
Collagen	Scar tissue, tumour	Masson's trichome

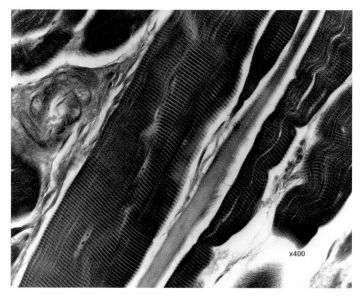

Figure 1.12 Phosphotungstic acid haematoxylin (PTAH).

stains after examining a slide (e.g. to confirm the presence of collagen and thus the diagnosis of a fibrosarcoma) and this will necessitate a protocol amendment (Keane, 2014). Alternatively, special stains may be incorporated into the study plan from its inception (e.g. when a sponsor requests special staining of all fat tissue in order to assess the effect of compounds such as peroxisome proliferator-activated receptor (PPARs) agonists on white and brown fat; Long et al., 2009).

1.7   Decalcification

Tissues which contain high levels of calcium slats such as bone and teeth are difficult to cut so they require decalcification with solutions such as formic acid to remove the calcium salts, soften the tissue and make it less brittle and easier to cut. Artefacts associated with excessive periods of tissue decalcification have been described (McInnes, 2011).

1.8   Immunohistochemistry

Immunohistochemistry (IHC) is used when H&E staining does not give sufficient information about the cell type of interest. Immunohistochemistry is requested by the pathologist when he or she wants definite confirmation or detection of a specific cell type. For instance, H&E staining cannot distinguish between T- and B-cells but IHC using antibodies against CD3 and CD19 can; thus, IHC is a method of detecting a molecule or epitope on a cell, in situ, in a tissue section, using an antibody to that molecule and a visible label.

Immunohistochemistry is useful in the identification of unknown tumours and in the detection of infectious agents. Toxicological pathologists may need to know the

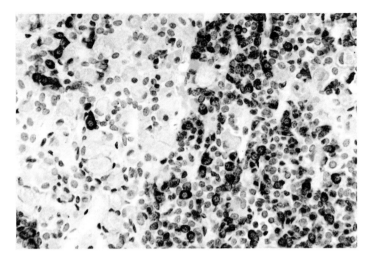

Figure 1.13 Growth-hormone IHC stains positive cells in pituitary.

exact identity of a cell that displays test article-related changes; markers such as glial fibrillary acid protein (GFAP) (which stains reactive astrocytes), synaptophysin and chromagranin (which stain neuroendocrine cells), growth hormone in pituitary cells (Figure 1.13) and LAMP-2(+) for hepatic phopholipidosis are thus all useful. Other common antibodies include proliferating cell nuclear antigen (PCNA), which stains cells in active proliferation (G1, G2, S and M phases), and the TUNEL and caspase 3 antibodies, which stain cells in apoptosis.

Fresh tissue samples frozen in OCT compound (Tissue Tek, UK) are optimal for IHC or in situ hybridisation, but formalin fixed tissue may also be used provided various techniques are applied to break down the formalin bonds (such as microwaving the tissue in citrate buffer, which is called 'antigen retrieval'). Cell morphology is generally poorer in fresh-frozen sections than in paraffin wax-embedded sections, but more antibodies tend to work on fresh-frozen tissue.

Mast cells are usually fairly easy to detect, due to their prominent granulation, but they can be highlighted, if required, using the histochemical stain toluidine blue or by using IHC with the antibody against CD117. Mononuclear immune cells such as macrophages and lymphocytes in rodents may be identified using a basic panel of IHC markers for macrophages (F4/80, Mac2), T-cells (CD3, CD4, CD8) and B-cells (CD19, CD23) (Ward et al., 2006).

Challenges in IHC include the fact that most antibodies are developed against human antigens, and thus toxicological pathologists are never sure whether a given antibody will work in rodent tissues (good positive and negative controls are essential); the difficulty of cutting suitable cryostat sections from frozen tissues; the problems involved in unmasking antigens (antigen retrieval) in formalin-fixed tissue; and high levels of background staining. Polyclonal antibodies bind to several different epitopes and are thus more sensitive but less specific than monoclonal antibodies, which bind to a single epitope.

1.9   Tissue Crossreactivity Studies

Tissue crossreactivity (TCR) studies are conducted in pharmaceutical environments and contract research organisations in order to check for off-target antigen binding and on-target binding in unusual tissues with new monoclonal antibodies intended for therapeutic use. The value of TCR studies lies in the prediction of toxicity that would not have been detected using in vivo pharmacology and toxicology studies in pharmacologically relevant animal models (Geoly, 2014). Generally, a complete set of frozen human and animal tissues is available for immunolabelling with the new antibody. The TCR assay by itself has variable correlation with toxicity and efficacy (Leach et al., 2010), and crossreactivity studies often involve a great deal of background staining in various animal species and in human tissues, which is difficult to interpret and may not always be the best method of determining the safety of a new antibody.

1.10   Electron Microscopy

Electron microscopy is used to confirm findings observed in H&E-stained tissue sections, such as hepatic hypertrophy (increase in peroxisomes and in endoplasmic reticulum), and in conditions such as phospholipidosis (lamellar bodies) (Figure 1.14). Transmission electron microscopy (TEM) produces two-dimensional (2D) photograph

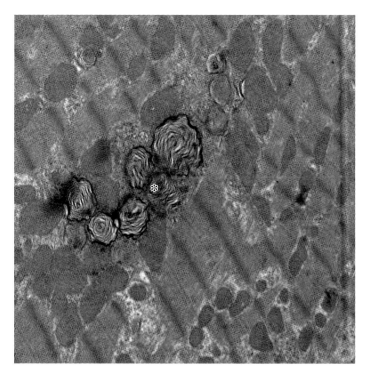

Figure 1.14 Phospholipidosis is characterised by lamellar bodies (*), which are visible in this electron micrograph of a rat heart.

of a thin section of tissue, whilst scanning electron microscopy (SEM) produces a three-dimensional (3D) photograph of the surface of a tissue. Generally, TEM is used because it provides greater resolution than does a light microscope and because it produces pictures of actual cells and their organelles and nuclei. In general TEM provides photographs which demonstrate cellular organelles, mitochondria, rough and smooth reticulum and lysosomes, phopholipidosis, inclusions and peroxisomes. In contrast SEM is useful in assessing the surface of the intestine in order to see bacterial attachment or the destruction of the villi. The disadvantages of electron microscopy include the cost, the use of particular fixatives, resin-embedding and the requirement for trained technicians to produce useful photographs. This has led to a recent reduction in its use.

1.11 In Situ Hybridisation

In situ hybridisation (ISH) can be used to detect RNA or DNA of interest in tissue sections. Polymerase chain reaction (PCR) and other gel/blot methods do not identify in which cell(s) the RNA has been increased (upregulated). Thus ISH is used to allow the pathologist to see which cells are producing a particular DNA or RNA product, because it provides the advantage that the architecture of the tissue is maintained (which is not the case in PCR).

In common with IHC, ISH requires different types of probe. Probes can be lengths of DNA or RNA. The RNA probes are often called 'riboprobes'. The most commonly used DNA probe in ISH is the oligonucleotide probe. Probes must have a sequence of bases (A,T,C,G) that is complementary to the mRNA of interest (i.e. if the base on the mRNA is C, the probe needs to be G). A number of companies will synthesise probes and label them. Binding of probe to target is termed 'hybridisation'. A label is used to visualise the probe in a particular cell or tissue.

Tissue microarrays are increasingly used to screen large numbers of tissues using either IHC or ISH in pharmaceutical environments. Tissue microarrays consist of multiple tissue samples embedded in a single wax block, used to produce a single glass slide containing many small samples of different tissues.

1.12 Laser Capture Microscopy

Areas of interest within a tissue section can be cut out of the section using a laser. The tissue can then be used in a number of subsequent techniques, such as amplification of mRNA in the area of interest by reverse transcriptase polymerase chain reaction (RT-PCR) or IHC.

1.13 Confocal Microscopy

Confocal microscopy provides high-resolution photographs of 2D or 3D fluorescent images. The confocal microscope has a pinhole and is efficient at rejecting out-of-focus fluorescent light. The practical effect of this is that an image comes from a thin section of the tissue. By scanning multiple thin sections in a sample, one can build up a very clean 3D image of the sample (Prasad et al., 2007).

1.14   Image Analysis

Image analysis provides quantitative assessments of cell number, cell size, lesion extent and the staining of a particular antibody in an organ, amongst other things. A computer program is used to generate the data, to which statistics can be applied. Further information about image analysis can be found in Scudamore (2014) and in Chapter 8 of this volume.

1.15   Digital Imaging

Digital imaging allows glass slides to be scanned and examined remotely online. This enables the slides to be viewed by multiple parties without the need for transport to different countries. Digital imaging is useful for quantitative analysis of cell and nuclear sizes and for examination of difficult lesions by multiple pathologists in different locations. It also facilitates the peer-review process.

1.16   Spermatocyte Analysis

Recognising the different cellular associations that make up the spermatogenic cycle within seminiferous tubules of the testis is essential to identifying when cell populations are missing. A further challenge in the male animal is how to recognise immaturity and distinguish it from degenerative conditions (Creasy, 2011). This is a problem in the dog and non-human primate. The use of the correct fixative to preserve the testes (Lanning et al., 2002; Latendresse et al., 2002) is essential for spermatogenesis testing. Examination of the different stages of the spermatogenic cycle in the testis of the rat and mouse using PAS-stained testis is a valuable technique that allows pathologists to evaluate spermatocyte development and identifies test article-related effects at various stages, if they exist (Creasy, 1997). Generally, most rodent studies will require the examination of a PAS-stained slide of the testis of the male rats and mice to ensure that no treatment-related findings have occurred in the male reproductive organs.

1.17   Good Laboratory Practice

Most studies in contract research organisations are conducted according to GLP (HHS, 1992). Medical devices, drugs and biological products intended for veterinary or human use are regulated by the EMA or FDA and must undergo preclinical testing for safety (generally in laboratory animals) before clinical trials in humans can begin. Adherence to GLP is intended to ensure that the quality and integrity of the data sent to the regulatory body can be relied upon and that no alteration of data have occurred! All regulations concerning GLP studies are comprehensive; they include standards for all personnel, facilities, equipment, test articles and records. Briefly, GLP requires the identification of individuals who fulfil responsibilities of management (as defined by GLP), as well as GLP-compliant facilities, equipment and materials. Training is essential for all personnel who are required to be GLP-compliant, and a quality-assurance (QA) department is required to inspect facilities and examine protocols and reports. The SOPs underpin GLP and should be in place for the histology, haematology, biochemistry and other laboratories. The SOPs should be available for the collection and identification of specimens, conduction of necropsies and the reception, identification, storage, and testing of control and test materials (e.g. serum, formalin-fixed tissues).

1.18 Inhalation Studies

Inhalation pathology concerns test article-related findings in the tissues of the upper respiratory system, which includes the larynx, nasal turbinates, trachea, tracheal bifurcation and associated lymph nodes. Rats, mice, dogs and non-human primates are commonly used in inhalation studies. Large structural differences have been described amongst these animal species in the anatomy of the nasal cavity, which cause variations in the intranasal airflow and deposition of particles, as well as total air volume. In addition, there are species differences in the turbinate sizes and consequent epithelial surface areas. Inhalation studies are more specialised than routine safety studies, and study personnel need to know the significance of animal inhalation lesions and background lesions in order to assess the risk of a particular inhaled compound for humans.

Four epithelial types are found in the nasal cavity/nasal turbinates (Monticello et al., 1990): squamous epithelium, transitional epithelium, respiratory epithelium and olfactory epithelium. Standardisation of nasal sections for histopathological evaluation is essential, so the four sections of the rat and mouse nasal turbinates are always cut in the same locations of the upper palate (Young, 1981). In rodents, the base of the epiglottis is an important predilection site for xenobiotic damage, where sensitive epithelium overlies the submucosal glands. Ultimately, cynomologus monkeys may be a better laboratory animal for testing human inhaled compounds, as their nasal structures more closely resemble those of humans than do those of rodents. Good laboratory practice ensures that raw data are uniform, consistent, reliable and reproducible (Keane, 2014).

1.19 Continuous-Infusion Studies

Continuous-infusion studies involve the use of medical pumps attached to the animal (generally in cloth pouches), which allow for the continuous infusion of a compound into a particular vein (via a catheter). They have a number of particular problems, including how to choose the species best suited to an indwelling catheter, the infusion technique, the surgical method used to attach the pump and insert the catheter, the chemical composition of the catheter material and the injected compound and the choice of blood vessel and its diameter, as well as how to keep the catheter open during the in-life period (Weber et al., 2011). Polyurethane catheters are preferred over polyethylene catheters, which can harden, become brittle and release small amounts of catheter substance into the circulation (Weber et al, 2011). A continuous-infusion study in the dog using the saphenous vein will not be sustained for more than 3 days, because the animal scratches the wounds and pulls the catheter out of the blood vessels (Weber et al., 2011).

The length of the jugular is often overestimated in rabbits and dogs, and the catheter extends into the ventricles of the heart, where it causes inflammation of the endocardium (Weber et al., 2011). Lesions such as chronic inflammation, fibrosis, foreign-body reactions to sutures (granuloma formation), abscess and thrombosis and haemorrhage are typical at the site of insertion of the catheter into the blood vessel (Weber et al., 2011) and should be distinguished from test article-related lesions. High incidences of abscesses and necrosis in the study animals suggest poor technique and make the study invalid.

1.20 Carcinogenicity

Carcinogenicity studies are generally conducted in mice and rats, which are exposed to test compounds for a period of approximately two years in order to establish whether they cause an increased incidence of tumours in the treated animals compared to non-treated controls. Carcinogenicity studies require large numbers of animals (generally 50 per sex group) in order to ensure that sufficient numbers will still be alive at the end of 104 weeks. Expected benefits of carcinogenicity studies include an increase in the safety of people exposed to chemicals that have passed toxicity tests, increased efficiency during the development of human pharmaceuticals and decreased wastage of animals, personnel and financial resources. Carcinogenicity studies are expensive and require excellent management and organisation skills amongst study personnel. In addition, statistical design issues (e.g. proper randomisation of animals, sample size considerations, dose selection and animal allocation issues, as well as the control of potentially confounding factors such as littermate and caging effects) will need to be addressed (Haseman, 1984). Since carcinogenicity studies take a long time and consume many resources, they should only be performed when human exposure warrants the collection of data from life-time studies in animals.

Recently, the poor predictability and correlation between treated rodents and humans in conjunction with their substantial animal-welfare and economic costs, have brought into question their purpose and future (Knight, 2007).

1.21 Biologicals

Biological medical products, or 'biologicals', are compounds manufactured in or derived from biological sources. This means that they are different to pharmaceutical products which are chemically produced, such as anti-inflammatory drugs. Examples of biologicals include living cells (such as stem cells), vaccines, gene therapy, recombinant proteins and blood and blood products. Biologicals can consist of sugars, proteins or nucleic acids, or they may be cells and tissues. Biologicals may be regulated through different methods than conventional chemical compounds.

1.22 The Pathology Report

Pathologists record their findings in a secure and validated computer data-capture program (such as Provantis), which allows for table generation and statistical analysis (see Chapter 2). Peer review is required for GLP-compliant studies and is conducted to produce consistent and reliable data and report (see Chapters 2 and 8). Once the study pathologist and the peer-reviewing pathologist have agreed on the findings, the study pathologist makes the necessary changes to the data, locks the data and generates a final set of tables and a final report. This report is then sent to the study director for incorporation into the final study report. Quality pathology reports are generated when motivated study personnel and the study pathologist communicate and work together (Keane, 2014).

The pathology report must be concise and accurate, and it should state the significance of its findings (i.e. what is and what is not test article-related). The macroscopic and histopathological findings should be interpreted in conjunction with the haematology and biochemistry results, as well as the organ weights and macroscopic findings should be correlated with histopathology findings if possible. A statement about the no-observed-adverse-effect level (NOAEL) should also be included (see Chapter 7). The pathology report should also identify any treatment related, unscheduled deaths which occurred before the study was terminated (Keane, 2014).

1.23 Conclusion

It is impossible to list all of the potential pathological techniques used by all pharmaceutical companies and contract research organisations. However, this chapter has attempted to cover some of the more common techniques. It should also provide useful information about how the glass slides used in toxicological pathology studies are made.

References

Braber, S., Verheijden, K.A., Henricks, P.A., Kraneveld, A.D. and Folkerts, G. (2010) A comparison of fixation methods on lung morphology in a murine model of emphysema. *American Journal of Physiology: Lung Cellular and Molecular Physiology*, 299, L843–51.

Connolly, H.M., Crary, J.L., McGoon, M.D., Hensrud, D.D., Edwards, B.S., Edwards, W.D. and Schaff, H.V. (1997) Valvular heart disease associated with fenfluramine-phentermine. *New England Journal of Medicine*, 337(9), 581–8.

Costa, S., García-Lestón, J., Coelho, M., Coelho, P., Costa, C., Silva, S., Porto, B., Laffon, B. and Teixeira, J.P. (2013) Cytogenetic and immunological effects associated with occupational formaldehyde exposure. *Journal of Toxicology and Environmental Health, Part A*, 76, 217–29.

Creasy, D.M. (1997) Evaluation of testicular toxicity in safety evaluation studies: the appropriate use of spermatogenic staging. *Toxicologic Pathology*, 25, 119–31.

Creasy, D.M. (2011) Reproduction of the rat, mouse, dog, non-human primate and minipig. In: McInnes, E.F. (ed.). *Background Lesions in Laboratory Animals*, Saunders, Edinburgh, pp. 101–22.

Fiette, L. and Slaoui, M. (2011) Necropsy and sampling procedures in rodents. *Methods in Molecular Biology*, 691, 39–67.

Geoly, F.J. (2014) Regulatory forum opinion piece: tissue cross-reactivity studies: what constitutes an adequate positive control and how do we report positive staining? *Toxicologic Pathology*, 42(6), 954–6.

Haseman, J.K. (1984) Statistical issues in the design, analysis and interpretation of animal carcinogenicity studies. *Environmental Health Perspectives*, 58, 385–92.

HHS. 1992. *Good Laboratory Practice (GLP) for Nonclinical Laboratory Studies, Department of Health and Human Services*, Washington, DC.

Keane, K. (2014) Histopathology in toxicity studies for study directors. In: Brock, W.J., Mouhno B. and Fu Lijie (ed.). *The role of the study director in nonclinical studies: Pharmaceuticals, chemicals, medical devices, and pesticides*, John Wiley & Sons, Inc. pp. 275–295.

Kittel, B., Ruehl-Fehlert, C., Morawietz, G., Klapwijk, J., Elwell, M.R., Lenz, B., O'Sullivan, M.G., Roth, D.R. and Wadsworth, P.F.; RITA Group; NACAD Group. (2004) Revised guides for organ sampling and trimming in rats and mice – Part 2. A joint publication of the RITA and NACAD groups. *Experimental and Toxicologic Pathology*, 55, 413–31.

Knight, A. (2007) Animal experiments scrutinised: systematic reviews demonstrate poor human clinical and toxicological utility. *ALTEX*, 24(4), 320–5.

Knoblaugh, S., Randolph Habecker, J. and Rath S. (2011) Necropsy and histology. In: Treuting, P.M. and Dintzis, S. (eds). *Comparative Anatomy and Histology: A Mouse and Human Atlas*, Elsevier, Amsterdam, pp. 15–41.

Lanning, L.L., Creasy, D.M., Chapin, R.E., Mann, P.C., Barlow, N.J., Regan, K.S. and Goodman, D.G. (2002) Recommended approaches for the evaluation of testicular and epididymal toxicity. *Toxicologic Pathology*, 30, 507–20.

Latendresse, J.R., Warbrittion, A.R., Jonassen, H. and Creasy, D.M. (2002) Fixation of testes and eyes using a modified Davidson's fluid: comparison with Bouin's fluid and conventional Davidson's fluid. *Toxicologic Pathology*, 30, 524–33.

Leach, M.W., Halpern, W.G., Johnson, C.W., Rojko, J.L., MacLachlan, T.K., Chan, C.M., Galbreath, E.J., Ndifor, A.M., Blanset, D.L., Polack, E. and Cavagnaro, J.A. (2010) Use of tissue cross-reactivity studies in the development of antibody-based biopharmaceuticals: history, experience, methodology, and future directions. *Toxicologic Pathology*, 38(7), 1138–66.

Long, G.G., Reynolds, V.L., Dochterman, L.W. and Ryan, T.E. (2009) Neoplastic and non-neoplastic changes in F-344 rats treated with Naveglitazar, a gamma-dominant PPAR alpha/gamma agonist. *Toxicologic Pathology*, 37, 741–53.

McInnes, E.F. (2011) Artifacts in histopathology. In: McInnes, E.F. (ed.). *Background Lesions in Laboratory Animals*, Saunders, Edinburgh, pp. 93–9.

McInnes, E.F., Rasmussen, L, Fung, P., Auld, A.M., Alvarez, L., Lawrence, D.A., Quinn, M.E., del Fierro, G.M., Vassallo, B.A. and Stevenson, R. (2011) Prevalence of viral, bacterial and parasitological diseases in rats and mice used in research environments in Australasia over a 5-y period. *Lab Animal*, 40, 341–50.

Michael, B., Yano, B., Sellers, R.S., Perry, R., Morton, D., Roome, N., Johnson, J.K., Schafer, K. and Pitsch, S. (2007) Evaluation of organ weights for rodent and non-rodent toxicity studies: a review of regulatory guidelines and a survey of current practices. *Toxicologic Pathology*, 35, 742–50.

Monticello, T.M., Morgan, K.T. and Uraih, L. (1990) Nonneoplastic nasal lesions in rats and mice. *Environmental Health Perspectives*, 85, 249–74.

Morawietz, G., Ruehl-Fehlert, C., Kittel, B., Bube, A., Keane, K., Halm, S., Heuser, A. and Hellmann, J.; RITA Group; NACAD Group. (2004) Revised guides for organ sampling and trimming in rats and mice – Part 3. A joint publication of the RITA and NACAD groups. *Experimental and Toxicologic Pathology*, 55, 433–49.

Pearson, G.R. and Logan, E.F. (1978) The rate of development of postmortem artefact in the small intestine of neonatal calves. *British Journal of Experimental Pathology*, 59, 178–82.

Prasad, V., Semwogerere, D. and Weeks, E.R. (2007) Confocal microscopy of colloids. *Journal of Physics: Condensed Matter*, 19, 113102.

Ruehl-Fehlert, C., Kittel, B., Morawietz, G., Klapwijk, J., Elwell, M.R., Lenz, B., O'Sullivan, M.G., Roth, D.R. and Wadsworth, P.F.; RITA Group; NACAD Group. (2003) Revised guides for organ sampling and trimming in rats and mice – Part 1. A joint publication of the RITA and NACAD groups. *Experimental and Toxicologic Pathology*, 55, 91–106.

Scudamore, C.L. 2014. Practical approaches to reviewing and recording pathology data. In: Scudamore, C.L. (ed.). *A Practical Guide to the Histology of the Mouse*, John Wiley & Sons, Chichester, pp. 25–42.

Sellers, R.S., Morton, D., Michael, B., Roome, N., Johnson, J.K., Yano, B.L., Perry, R. and Schafer, K. (2007) Society of Toxicologic Pathology position paper: organ weight recommendations for toxicology studies. *Toxicologic Pathology*, 35, 751–5.

Seymour, R., Ichiki, T., Mikaelian, I., Boggess, D., Silva, K.A. and Sundberg, J.P. (2004) Necropsy methods. In: Hedrich, H. and Bullock, G. (ed.). *The Laboratory Mouse* (eds), Elsevier, New York, pp. 495–517.

Suvarna, S.K. and Ansary, M.A. (2001) Histopathology and the 'third great lie'. When is an image not a scientifically authentic image? *Histopathology*, 39, 441–6.

Tannenbaum, J. and Bennett, B.T. (2015) Russell and Burch's 3Rs then and now: the need for clarity in definition and purpose. *Journal of the American Association of Lab Animal Science*, 54, 120–32.

Ward, J.M., Erexson, C.R., Faucette, L.J., Foley, J.F., Dijkstra, C. and Cattoretti, G. (2006) Immunohistochemical markers for the rodent immune system. *Toxicologic Pathology*, 34, 616–30.

Weber, K., Mowat, V., Hartmann, E., Razinger, T., Chevalier, H.J., Blumbach, K., Green, O.P., Kaiser, S., Corney, S., Jackson, A. and Casadesus, A. (2011) Pathology in continuous infusion studies in rodents and non-rodents and ITO (infusion technology organisation) – recommended protocol for tissue sampling and terminology for procedure-related lesions. *Journal of Toxicologic Pathology*, 24, 113–24.

Young, J.T. (1981) Histopathological examination of the rat nasal cavity. *Fundamental and Applied Toxicology*, 1, 309–12.

2

Recording Pathology Data

Cheryl L. Scudamore

MRC Harwell, Harwell Science and Innovation Campus, Oxfordshire, UK

Learning Objectives

- Understand what constitutes a pathology finding.
- Understand how pathology findings can be quantified.
- Understand how a pathology tem is created in terms of location, distribution and chronicity.
- Understand historical control data and peer review.

In pharmaceutical companies and associated contract research organisations (CROs), pathology data are usually generated by a toxicologic pathologist. Historically, toxicologic pathologists in the pharmaceutical industry have come from a number of scientific backgrounds, with different primary degrees in biological sciences, medicine, dentistry and veterinary medicine. Currently, the majority of toxicologic pathologists are veterinary graduates, usually with postgraduate training and qualifications in pathology (Bolon et al., 2010; van Tongeren et al., 2011). As well as recognising and recording pathological findings, pathologists should also be able to interpret these findings in the context of macroscopic observations made in-life and at necropsy, changes in organ weights and other available pathology data (e.g. clinical biochemistry and haematology changes). All of this information should be interpreted in a study pathologist's report, taking into consideration what is already known about the compound, biopharmaceutical, gene or infectious agent under study (Crissman et al., 2004).

Pathologists often have multiple roles within an organisation, acting as study pathologists, project team members, researchers and managers. Whilst within the pharmaceutical industry the process of generating regulatory pathology (safety) data from necropsy, via slide production and through interpretation by a qualified experienced pathologist, is usually well regulated by Good Laboratory Practice (GLP) or an equivalent quality framework, this is not always the case in academia. It is therefore important when reviewing 'discovery' research data and published literature to be aware that the pathology may not have been reported by a qualified individual and that the terminology may vary from that used in toxicology studies (Cardiff et al., 2008).

Pathology for Toxicologists: Principles and Practices of Laboratory Animal Pathology for Study Personnel,
First Edition. Edited by Elizabeth McInnes.

2.1   What is a Pathology Finding?

Pathologists often refer to their results as 'findings', rather than 'lesions'. A 'finding' is something that a pathologist considers worth recording and may or may not be a pathological lesion.

1) Pathologists may record 'normal' alterations and variations in tissue morphology, because these changes may be modified in treated animals. Examples of normal changes which might be recorded include:

   - stage of oestrous cycle in females;
   - presence/degree of extramedullary haematopoiesis in rodent spleens;
   - presence of immature/pubescent reproductive organs, especially in studies involving large animal species (dogs and non-human primates).

2) Pathologists may also note pathology that is incidental and which may or may not be specifically related to the study. For example, trauma related to fighting or accidental injuries from caging may occur in any study, whereas injuries related to gavage in rodents or cystitis following bladder catheterisation in dogs may be related to study procedures rather than an effect of treatment.

3) Tissues from all species will show spontaneous or background incidental findings that are recognised and expected variations for a given strain or species (McInnes, 2012). Pathologists will generally record these, because increases or decreases in the incidence of these lesions may be associated with a treatment-related effect. For example, inflammatory cell foci are common in many organs, but particularly the liver (Foster, 2005), and increased foci may be associated with minimal levels of hepatotoxicity, whereas decreased foci may be seen if a compound has anti-inflammatory properties.

4) Pathologists may, finally, note lesions that are uniquely induced by the experimental protocol itself. These may be related to the compound under test in a toxicity study, or they may be the result of a genetic modification in a study involving genetically altered animals.

2.2   Standardisation of Pathology Findings

Pathology is an observational science, and it is therefore, by definition, difficult to ensure exact standardisation in the recording of findings between different observers. Two major techniques are used to enhance the standardisation and reproducibility of pathology findings: semiquantitative analysis of lesion severity and harmonised terminology nomenclatures.

2.2.1   Semiquantitative Analysis

Traditionally, diagnostic pathology employs qualitative narrative reports to record what the pathologist sees down the microscope. A well-written qualitative description provides a lot of information about the morphological changes present in a tissue and can be very useful in creating a record of a novel induced lesion, which can subsequently be recognised by other pathologists. However, as with any other scientific observations,

Table 2.1 Comparison of data types.

Qualitative	Semiquantitative	Quantitative
Description of morphological changes seen in tissue	Extent or severity of lesions divided into a number of discrete scores or grades	Measurement of numbers of cells, lesions and areas affected
Useful for recording novel induced lesions	Useful for rapid data recording, where a yes/no or ranked answer is sufficient	Useful where precise data are required and in separating subtle differences
Subjective	Subjective	Objective
Time-consuming	Faster	Faster for individual parameters, once image-analysis system is trained
		Slower if used to measure whole range of possible lesions in a study
Difficult to analyse	Can be analysed	Can be analysed statistically
Difficult to compare groups statistically	Allows statistical comparison using nonparametric tests	
Relies on experience of operator	Relies on experience of operator	Less reliant on experience

qualitative data are difficult to compare, subject to variation in style between individual analysts (pathologists) and hard to analyse statistically.

For many scientific parameters, quantitative (numerical) data are the expected output, but for pathology analysis, quantitation involves a significant input of labour and human intervention, in order to train computer systems and enable image-analysis tools to be used. In practice, semiquantitative analysis is almost universally used for high-throughput, non-neoplastic pathology studies, including toxicology studies, as it provides sufficiently reproducible data when used by experienced pathologists (Table 2.1). Comparison between semiquantitative and quantitative methods has shown that the overall conclusions are usually the same when analysing pathology data (Shackleford et al., 2002; Von Bartheld, 2002).

Neoplasms (i.e. tumours) are not normally graded semiquantitatively in terms of severity, but are usually recorded as being 'present' and categorised based on morphological criteria and biological behaviour as 'benign' or 'malignant' or as 'fatal' or 'nonfatal', to allow for Peto analysis (Peto et al., 1980).

Semiquantitative analysis involves allocating a grade or score to a lesion, based on its extent or severity. Ordinal grades or scores usually extend from 0 to between 3 and 5. Higher numbers of categories tend to be associated with less reproducibility, as it is hard for an individual to distinguish between and remember minor differences between grades. The system of scoring may be based on an approximately linear or nonlinear approach; some examples of descriptors for different grades are given in Table 2.2. A nonlinear approach is often preferred, as it allows background lesions to be acknowledged without overemphasising their significance (i.e. the assumption is that if a lesion is a background finding, it should not affect the tissue or organ function, and therefore should be present at a low grade at most; Mann et al., 2013).

Table 2.2 Examples of linear and nonlinear grading schemes that could be used for semiquantitative analysis of non-neoplastic lesions in laboratory animal tissues. *Source*: Scudamore (2014). Reprinted with permision from Wiley-Blackwell.

Linear		Nonlinear	
Grade	Description	Grade	Description
0	NAD (WNL): No change recorded	0	NAD (WNL): No change recorded
1	Minimal: 0–20% of tissue affected by change	1	Minimal: The least change that is visible on light microscopy at ×20; small, focal or affecting <10% of tissue
2	Mild (slight): 21–40% of tissue affected by change	2	Mild (slight): Change is readily detected but not a major feature; may involve multifocal small lesions or affect <20% of tissue; may still be within background appearance for the species
3	Moderate: 41–60% of tissue affected by change	3	Moderate: Change is more extensive or involves more foci (e.g. seen in every ×20 field), beyond the usual background for the lesion in the species; may start to have relevance for organ function and may correlate with other changes (e.g. increased organ weight)
4	Moderately severe (moderately marked): 61–80% of tissue affected by change	4	Moderately severe (moderately marked): As for 3, but more of the tissue is affected (e.g. up to 75%); likely to have relevance for tissue/organ function
5	Severe (marked): 81–100% of tissue affected by change	5	Severe (marked): Virtually the whole tissue is affected by the change, which is likely to be functionally relevant/detrimental

NAD, nothing abnormal detected; WNL, within normal limits.

Different terms may be used for grade '0', and it is important to know what the pathologist means when using them, as this can indicate the lower threshold of observation that they are using.

- NAD: 'Nothing abnormal detected'. Suggests that the particular finding is not present in the tissue.
- WNL: 'Within normal limits'. Suggests that very low levels of a finding may be present in a tissue but have not been recorded.

2.2.2 Nomenclature/Controlled Terminology

To ensure that pathology data are analysable, searchable and comparable between studies, within an organisation and between organisations, standardised terminology should be used. Controlled terminology also facilitates data capture when using computer systems and sharing of information with regulatory bodies (e.g. via SEND (Standard for Exchange of Nonclinical Data), an implementation of the CDISC (Clinical Data Interchange Standards Consortium: http://www.cdisc.org/send) Standard Data Tabulation Model (SDTM) for sharing nonclinical data).

Whilst some organisations may have their own in-house glossary of terms, there is an ongoing global collaborative process to produce a harmonised nomenclature: the International Harmonization of Nomenclature and Diagnostic (InHAND) criteria (Mann et al., 2013). This nomenclature will apply initially to rats and mice, but will ultimately extend to include large animal species. The InHAND initiative aims to publish diagnostic criteria and differential diagnoses for neoplastic (proliferative) and non-neoplastic findings. Neoplastic findings are often easier to record consistently, as they need to match specific diagnostic criteria, whereas non-neoplastic lesions are open to more interpretation.

Non-neoplastic lesions in the InHAND criteria are given distinct terminology, which, where possible, is descriptive rather than diagnostic (i.e. it does not imply a specific cause for a lesion). In some cases, where tissue injury results in ongoing damage and simultaneous repair, or where a lesion is made up of a constellation of changes, complex terms may be used which acknowledge this (e.g. 'reversible cell damage/regeneration' or 'nephropathy').

In addition to the actual term given to a lesion, other modifiers may be added to give more specificity to the pathology data (Figure 2.1):

- Location: The specific location of injury within a tissue may be important, so this can be specified (e.g. in the kidney, the pathology may affect the glomeruli, tubules or interstitial tissue between the tubules and glomeruli).
- Distribution: As with macroscopic lesions (see Chapter 1), lesions may affect the whole tissue (i.e. be diffuse) or only a portion (i.e. be focal).
- Chronicity: Sometimes, a term may be added to indicate the possible age of a lesion, ranging from 'subacute' (very recent) to 'acute' or 'chronic' (longstanding). Whilst these terms may be helpful in some circumstances (e.g. a subacute lesion found at the end of a long-term study may not be related to treatment), the timings associated with them are not fixed and are not necessarily the same for a given tissue, so they should be viewed with caution.
- Process terms: Sometimes, terms are used which indicate the type of process seen; these are often used for inflammation, but sometimes for other changes, too (e.g. fibrosis). Inflammation is often described based on the predominant inflammatory cell present (e.g. 'neutrophilic' (purulent), 'lymphoid', 'histiocytic' (granulomatous)).

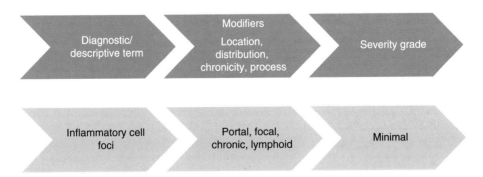

Figure 2.1 How a pathology term is created.

2.2.3   Ontological Approach

Whilst not widely used in toxicologic pathology, another means of controlling terminology is to use an ontological approach. Ontologies categorise the relationships between objects or terms on a hierarchical basis, creating, in this case, a taxonomy of pathology terms. Ontologies are used in many areas of medicine and pathology to enable integration of interdisciplinary and translational concepts and findings. One of the main interests in laboratory animals using this approach is to allow mapping of novel pathologies in genetically engineered animals to known human disease entities (Schofield et al., 2013). This approach can be computerised, enhancing data retrieval and computational modelling. It is important to be aware that this approach may be used when assessing pathology data from the wider academic community, where some compounds or biologicals will be discovered before being transferred to a pharmaceutical company for development.

2.3   'Inconsistencies' in Pathology Recording

An experienced pathologist will normally be consistent within a given study, and the use of controlled nomenclature with clear diagnostic criteria aids consistency between pathologists, especially with regards to proliferative lesions, which can clearly be mapped against a given set of features. Overall inexperience or specific inexperience with a given species or set of lesions can result in reduced consistency in terms of lesion classification and grading, but the process of informal peer review (see Section 2.6) and continual professional development is helpful in reducing this.

There are some specific causes of inconsistency that it is helpful to understand in order to appreciate their implications.

2.3.1   Diagnostic Drift

This can occur in a large study (e.g. a lifetime carcinogenicity bioassay, where tissues are reviewed over months or even years). The pathologist may start to over- or under-record certain background lesions with respect to their starting point over time. It is therefore important to carefully analyse incidence tables at the end of a study (and to review tissues, if necessary) if any anomalies are apparent. If data are reviewed partway through a study, it is important to remember that they may not represent the pathologist's final opinion once he or she has recognised and corrected for diagnostic drift.

2.3.2   Thresholds

No tissue from an animal will be completely free of 'findings', and if a pathologist looks hard enough, he or she will find something to record (e.g. small foci of inflammatory cells in liver). All pathologists will thus use some degree of threshold, which may be influenced by time pressure, speed of review or the use of only the lowest-power microscope objective. The terms 'NAD' and 'WNL' are used when a pathologist has not seen anything above his or her personal threshold. Thresholds will vary depending on the experience of the pathologist, type of study, number of animals and company or group policy.

The overall result of a study may not be affected by the levels of threshold that an individual pathologist uses, but differences in the data may lead to issues when comparing between studies read by different pathologists. Pathologists with a higher recording threshold (who generally do not record a lesion unless it is quite prominent) will generally have more animals with 0 or NAD findings.

2.3.3   Lumping versus Splitting

As well as having different thresholds for recording observations, pathologists may also choose to 'lump' or 'split' findings, either all the time or under certain circumstances. Lumping is the practice of using a single overarching term to cover a lesion made up of multiple individual findings, whilst splitting is the practice of breaking up a complex lesion into its component parts and recording these as individual findings. Lumping terms means there will be fewer findings recorded for a study and allows tracking of complex lesions with time, but it makes the assumption that the individual findings that make up the complex term are related only to that particular pathogenesis. A classic example of this dilemma is the recording of chronic progressive nephropathy (CPN), a common background finding in most rat strains (see Chapter 4). CPN is made up of a group of related findings: tubular basophilia, tubular dilatation and protein casts, interstitial inflammatory infiltrate and glomerular sclerosis. In the early stage of the pathology, only one or two of these findings may be present (often, minimal levels of tubular basophilia are seen), but as the lesion progresses, all of them will be present. This leads to significant differences in how the data look when recorded by a 'lumper' versus a 'splitter' (Figure 2.2).

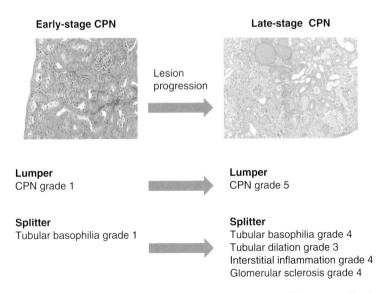

Figure 2.2 Chronic progressive nephropathy (CPN) is a common and well-recognised background finding in rats. The lesion progresses from occasional basophilic tubules in the early stages to multiple different morphological findings in the later (chronic) stages. A pathologist who lumps this lesion will record different grades of CPN throughout the time course of the lesion whereas a splitter will describe the different findings seen at each stage.

The two approaches both have benefits and problems. Lumping allows one to easily assess the progression of a lesion with time (i.e. as study length increases), but the individual findings are nonspecific, so it can be difficult to be certain that the early changes (which may consist of only one or two of the group of findings) are not due to another cause or part of another lesion. The combination of changes may be more specific for a well-recognised lesion, but if the minimal changes are split, it is hard to reconstruct a combination term from the data once they are complete.

These different approaches explain some of the apparent differences between pathologists' data sets, and will also impact on historical control incidences.

2.4 Blind Review

Blind review, where tissues are examined without prior knowledge of group or protocol information, is sometimes suggested as a means of reducing inconsistency and bias in pathology recording. However, it is not recommended for expert pathologists reading toxicity safety studies, for a number of reasons:

- It may reduce the likelihood of identifying any changes, particularly those which represent a subtle variation from 'normal' background findings.
- It may not be unbiased, because if there are obvious treatment-related tissue changes, these will effectively 'unblind' the examination.
- It does not make best use of the pathologist's knowledge, interpretive skills and experience (e.g. the ability to correlate clinical pathology and clinical signs information with a specific histopathological lesion).
- It slows the process (e.g. if slides have to be coded and decoded prior to reading) and adds extra cost to the study.

Whilst blind review is not recommended in best-practice guidance (Crissman et al., 2004; Neef et al., 2012) for global reading of histopathology studies in safety studies, there are circumstances in which it may be useful. Once a pathologist has reviewed an entire study in an unblinded fashion, it may be helpful to do a blind review of just the target organs, particularly when there is a subtle difference between control and treated animals (Holland and Holland 2011).

Blind review of slides is more commonly used in experimental and academic studies. In these cases, the reviewer may not be a trained pathologist. Researchers may also have a vested career interest in the outcome of a study, and blind review helps mitigate this bias.

2.5 Historical Control Data: Pros and Cons

Historical control data (HCD) usually comprise incidences of pathological lesions in control or wild-type animals collated from a number of studies, usually at a given institution or laboratory. There is no strict definition of what these data sets consist of, so it is important to understand the source and what is and is not included in any data sets used for comparison.

Ideally, concurrent controls, producing the expected results, should be available for all studies, but this is not always the case. Historical control pathology data can be useful even when concurrent controls are present; for example, in helping understand the significance in carcinogenicity studies of (Keenan et al., 2009):

- a rare tumour in a treated group;
- unexpected tumour incidences in controls;
- a marginal increase in the incidence or severity of proliferative lesions compared to a concurrent control.

HCD in these instances may add to the weight of evidence being used to assess carcinogenicity.

Sometimes, by chance, background lesions occur only in the high-dose treatment group, with no similar lesions in the control group. In this situation, HCD can be used to evaluate the expected 'normal' incidence of the finding in control animals in the facility.

HCD can also be used to benchmark changes in the incidences of lesions in a colony of animals over a period of time or to control for small group sizes in research and discovery studies or where there are no appropriate controls available.

It is important to critically evaluate all available HCD and to recognise the potential factors that can limit their usefulness:

- Variations due to what the pathologist records (i.e. individual variation and changes in nomenclature) can lead to variable results in the data (e.g. different terms being used for the same lesion in different studies). For instance, 'cystic degeneration of lymph nodes' is also known as 'sinus dilatation', 'lymphangiectasia', 'cystic ectasia', 'lymphatic sinus ectasia', 'lymphangiectasis' and 'lymphatic cysts'.
- Variations due to changes in animal substrain and genetic drift can lead to changes in the incidence of proliferative and nonproliferative lesions.
- Changes in husbandry/feed/health status can produce different results (e.g. reduction of chronic nephropathy in certain rat strains under conditions of dietary restriction).
- Changes in pathology technical practice may mean that different tissues are collected or prepared differently, which can have the effect of disguising or revealing lesions.

Other practical factors that need to be considered are whether the data are recorded as incidences, percentages or means and whether there is enough information to convert them into something that can be compared with the current study. Also, data that are accessible online may be easier (less time-consuming) to use and reanalyse than those held on paper.

An awareness of these limitations and the use of some additional precautions will ensure that HCD can be used appropriately and accurately. The following suggestions can help increase the value of HCD:

- Reduce the effects of genetic drift by only using data collected from animals in the last 2–7 years.
- Choose data from animals with similar husbandry and pathology practices. If this information is not available, treat the data with caution.

- Use data from studies where the pathology was peer reviewed, as they are likely to be more consistent and reliable.
- If necessary, have the actual slides reread by a pathologist (e.g. if nomenclature changes or if individual pathologist variation may be a problem).

2.6   The Use of Peer Review in Pathology

Peer review is not strictly required by GLP, but it is expected by most regulatory bodies. Guidance on its use has been provided by the Organisation for Economic Co-operation and Development (OECD, 2014). The reporting of histopathology data depends some-what uniquely on the 'opinion' of the study pathologist, and for this reason most organi-sations will have an internal (contemporaneous) peer-review process, where a percentage of slides are reassessed by a second pathologist. This helps ensure the quality and repro-ducibility of data and promotes the use of best practice. Peer review can also help identify and correct diagnostic drift where a study has been read over a long period of time. The study pathologist still retains responsibility for the pathology report. In this context, peer review may also form part of the ongoing training of pathologists, as the original and reviewing pathologist may learn from each other during the process (Mann and Hardisty, 2013a).

When a study is performed at a CRO on behalf of an external sponsor, the sponsor may conduct a retrospective peer review. A sponsor pathologist might have greater knowledge of the expected effects of a given compound, due to having access to better information about the compound's PK/PD and to studies conducted at different organi-sations. In addition, many CROs anonymise the compound name, so the study patholo-gist and study personnel will have limited information about the compound and may be ignorant of important factors which a pathologist from the company producing the compound would be aware of.

At whatever stage a peer review is done, it is not an attempt to reread and rereport a study. Normally, all the tissues from a percentage of study animals are reviewed, in addi-tion to all suspected target organs, neoplasms and proliferative lesions.

Pathology working groups (PWGs) are a specialised form of peer review that may be used when findings are controversial or pivotal in a development plan (Mann and Hardisty, 2013b). The PWGs are usually composed of a group of experts, who are asked to evaluate very specific lesions or groups of lesions from a given study, and to provide an unbiased assessment of the lesions and their relevance.

References

Bolon, B., Barale-Thomas, E., Bradley, A., Ettlin, R.A., Franchi, C.A., George, C., Giusti, A.M., Hall, R., Jacobsen, M., Konishi, Y., Ledieu, D., Morton, D., Park, J.H., Scudamore, C.L., Tsuda, H., Vijayasarathi, S.K. and Wijnands, M.V. (2010) International recommendations for training future toxicologic pathologists participating in regulatory-type, nonclinical toxicity studies. *Toxicologic Pathology*, 38, 984–92.

Cardiff, R.D., Ward, J.M. and Barthold, S.W. (2008) 'One medicine – one pathology': are veterinary and human pathology prepared? *Laboratory Investigation*, 88, 18–26.

Crissman, J.W., Goodman, D.G., Hildebrandt, P.K., Maronpot, R.R., Prater, D.A., Riley, J.H., Seaman, W.J. and Thake, D.C. (2004) Best practices guideline: toxicologic histopathology. *Toxicologic Pathology*, 32, 126–31.

Foster, J.R. (2005) Spontaneous and drug-induced hepatic pathology of the laboratory beagle dog, the cynomolgus macaque and the marmoset. *Toxicologic Pathology*, 33, 63–74.

Holland, T. and Holland, C. (2011) Analysis of unbiased histopathology data from rodent toxicity studies (or, are these groups different enough to ascribe it to treatment?). *Toxicologic Pathology*, 39, 569–75.

Keenan, C., Elmore, S., Francke-Carroll, S., Kemp, R., Kerlin, R., Peddada, S., Pletcher, J., Rinke, M., Scmidt, P.S., Taylor, I. and Wolf, D.C. (2009) Best practices for use of historical control data of proliferative rodent lesions. *Toxicologic Pathology*, 37, 679–93.

Mann, P.C., Vahle, J., Keenan, C.M., Baker, J.F., Bradley, A.E., Goodman, D.G., Harada, T., Herbert, R., Kaufmann, W., Kellner, R., Nolte, T., Rittinghausen, S. and Tanaka, T. (2013) International harmonization of toxicologic pathology nomenclature: an overview and review of basic principles. *Toxicologic Pathology*, 40(4 Suppl.): 7S–13S.

Mann, P.C. and Hardisty, J.F. (2013a) Peer review and pathology working groups. In: Haschek, W.M., Rousseaux, C.G., Wallig, M.A., Bolon, B., Ochoa, R. and Mahler, B.W. (eds). *Toxicologic Pathology*, 3rd edn, Elsevier, New York, pp. 551–64.

Mann, P.C. and Hardisty, J.F. (2013b) Pathology working groups. *Toxicologic Pathology*, 42, 283–4.

McInnes, E.F. (2012) Preface. In: McInnes, E.F. (ed.). Background Lesions in Laboratory Animals: A Colour Atlas, Saunders Elsevier, Amsterdam, p. vi.

Neef, N., Nikula, K., Francke-Carroll, S. and Boone, L. (2012) Regulatory forum opinion piece: blind reading of histopathology slides in general toxicology studies. *Toxicologic Pathology*, 40, 697–9.

OECD. (2014) OECD Series on Principles of Good Laboratory Practice and Compliance Monitoring. Number 16: Advisory Document of the Working Group on Good Laboratory Practice – Guidance on the GLP Requirements for Peer Review of Histopathology, OECD Publishing, Paris. Available from: http://www.oecd.org/officialdocuments/publicdisplay documentpdf/?cote=env/jm/mono(2014)30&doclanguage=en (last accessed July 29, 2016).

Peto, R., Pike, M.C., Day, N.E., Gray, R.G., Lee, P.N., Parish, S., Peto, J., Richards, S. and Wahrendorf, J. (1980) Guidelines for simple, sensitive significance tests for carcinogenic effects in long-term animal experiments. *IARC Monographs on Evaluation of the Carcinogenic Risk of Chemicals to Humans. Supplement*, (2 Suppl.), 311–426.

Schofield, P.N., Sundberg, J.P., Sundberg, B., McKerlie, C. and Gkoutos, G.V. (2013) The mouse pathology ontology, MPATH; structure and applications. *Journal of Biomedical Semantics*, 4, 18.

Scudamore, C.L. (2014) Practical approaches to reviewing and recording pathology data. In: A Practical Guide to Histology of the Mouse, John Wiley & Sons, Chichester.

Shackleford, C., Long, G., Wolf, J., Okerberg, C. and Herbert, R. (2002) Qualitative and quantitative analysis of nonneoplastic lesions in toxicology studies. *Toxicologic Pathology*, 30, 93–6.

van Tongeren, S., Fagerland, J.A., Conner, M.W., Diegel, K., Donnelly, K., Grubor, B., Lopez-Martinez, A., Bolliger, A.P., Sharma, A., Tannehill-Gregg, S., Turner, P.V. and Wancket, L.M. (2011) The role of the toxicologic pathologist in the biopharmaceutical industry. *International Journal of Toxicology*, 30, 568–82.

Von Bartheld, C.S. (2002) Counting particles in tissue sections: choices of methods and importance of calibration to minimise biases. *Histology and Histopathology*, 17, 639–48.

3

General Pathology and the Terminology of Basic Pathology

Elizabeth McInnes

Cerberus Sciences, Thebarton, SA, Australia

Learning Objectives

- Understand common pathological changes and terminology (e.g. 'necrosis', 'inflammation', 'circulatory changes', 'neoplasia', 'repair').
- Understand the chronicity of lesions (i.e. acute/chronic inflammation).
- Learn how to differentiate between related pathologies (e.g. hyperplasia, hypertrophy, neoplasia).
- Understand immune responses.

Pathology is divided into general pathology, which is the study of the actual processes of disease, and systemic pathology, which is the study of diseases or lesions within a specific body tissue (such as the liver). Toxicological pathology is concerned with the lesions caused by the administration in laboratory animals of new compounds intended for human therapeutic use.

Study personnel will find the language that pathologists use perplexing and esoteric. This is not a reflection on the ability of the study personnel! Although difficult for non-pathologists to understand, the complex language of pathology is used to convey precisely the observations made by the pathologist. In general, necrosis, degeneration and vacuolation are given the suffix '-opathy' (a general term indicating a pathologic necrotic or degenerative condition in a specific organ). This is preceded by the organ name (generally in Greek), indicating the tissue or organ affected. An example is 'nephropathy', which indicates necrosis or degeneration in the kidney ('nephros'). Qualifiers such as distribution, duration, severity and type of cell involved are used to 'build a diagnosis'; for example, 'multifocal, chronic, severe nephropathy'.

3.1 Cellular Responses to Insults

Pathology revolves around different types of cell injury. Severe cell injury is not difficult to recognise (Figure 3.1), and cell injury is reversible up to a certain point (Kumar et al., 2010a), but if the damage persists, then the cell undergoes irreversible injury and cell

Pathology for Toxicologists: Principles and Practices of Laboratory Animal Pathology for Study Personnel, First Edition. Edited by Elizabeth McInnes.

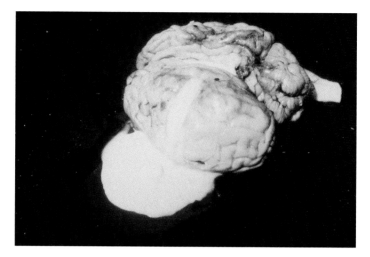

Figure 3.1 Severe cell injury is not difficult to recognise. Here, the pus exuding from this ruminant brain clearly indicates that the tissue is not normal. This is called 'purulent meningitis'.

death (necrosis). Causes of cell damage and injury include reduced oxygen supply, oxygen-derived free radicals, physical agents, chemical agents, toxins, infectious agents (e.g. bacteria and viruses), hypersensitivity and immune reactions.

Reversible cell damage (degeneration) may be slightly more difficult to recognise at necropsy than necrosis. The principle recognisable features of reversible damage are cell swelling and fatty change, both of which are present in the fatty liver illustrated in Figure 3.2. Swelling of the endoplasmic reticulum and mitochondria, as well as myelin figures (whorled masses) and membrane blebs, are seen in electron microscope photographs of reversible cell damage (Kumar et al., 2010a). Reversible cell injury often involves the accumulation of substances within the cytoplasm of a cell, and is characterised by reduced energy in the cell and cell swelling. Accumulated substances are

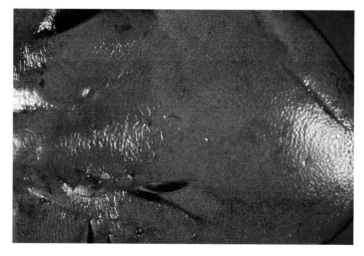

Figure 3.2 Yellow fatty change in an enlarged ruminant liver.

produced by the body; these include fluid (termed 'hydropic change'), lipid (fat) (termed 'lipid', 'lipidosis' or 'fatty change'; see Figure 3.2) and pigments produced during cellular breakdown ('wear-and-tear pigments'), such as haemosiderin (a golden-brown product produced when red blood cells are broken down), bile (a greenish-yellow product) and lipofuscin (a golden-brown pigment). Lipofuscin is found in liver cells, cardiac muscle cells, adrenal cortex, testis, ovary and neurons in the brain. Bile pigments (bilirubin), haemosiderin (brown) and melanin (black) are endogenous pigments that can accumulate in cells. Glycogen is present in liver cells, and may accumulate in the kidney and liver in diabetes mellitus. Cholesterol may accumulate in foamy macrophages in the lung. Techniques for distinguishing between the different types of pigment include the use of special stains such as Sudan black B (fat), Schmorl's and Long Ziehl-Neelsen technique for lipofuscin, PAS with diastate for glycogen, Perl's Prussian blue for haemosiderin and Masson-Fontana for melanin (see Table 1.1).

Necrosis is easy to recognise at necropsy, generally presenting as a yellow to whitish discolouration of tissues, often with a soft or dry consistency. Necrosis is the death of cells and tissues in an animal that is still alive; the cells will show nuclear breakdown, breakdown of plasma membranes and leakage of cell contents under the light microscope (Golstein and Kroemer, 2007). One of the most dramatic and common causes of necrosis is ischaemic damage, which occurs when the blood supply to an organ or part of an organ is reduced or obstructed. Ischaemia is defined as an incident that prevents delivery of substrates and oxygen to tissues. A localised area of ischaemia is called an 'infarct' (Figure 3.3), which generally displays a sharp demarcation between normal and necrotic tissue. In the kidney, infarcts are often wedge-shaped, because this corresponds with the area previously supplied by a single arteriole that has become obstructed.

There are a number of different types of necrosis. Coagulative necrosis (Figure 3.4) occurs in solid organs such as the liver and kidney and appears pale white with a sharp demarcation between the necrotic tissue and the normal vascularised tissue. Liquefactive necrosis occurs when enzymes cause the breakdown of dead tissue to form a viscous

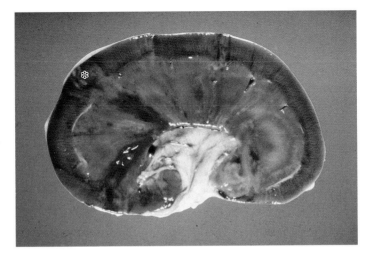

Figure 3.3 Infarcts (*) in a dog kidney, showing a sharp demarcation between normal (red-brown) tissue and necrotic (yellow) tissue.

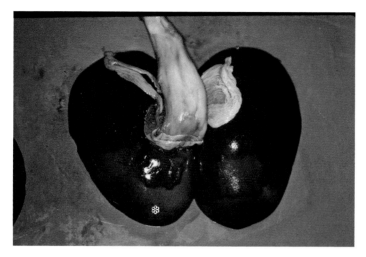

Figure 3.4 Coagulative necrosis (*) in a ruminant kidney.

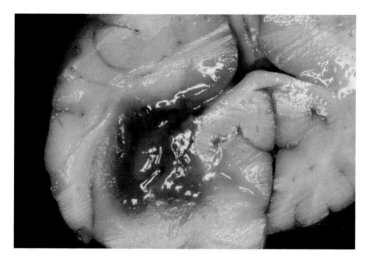

Figure 3.5 Liquefactive necrosis in a ruminant brain.

fluid; it is often seen in the brain, where it is called 'malacia' (Figure 3.5). Caseous necrosis is generally encountered in tuberculous lesions and involves cheese-like necrotic material (Figure 3.6). Fat necrosis involves fatty tissue becoming hard and often mineralised (calcium accumulates within tissues) due to the release of lipases (often because of pancreatitis), which digest fat; it is often seen in the mesenteric fat tissue (Figure 3.7). Gangrene is a variant of coagulative necrosis; it often involves extremities such as the tail or feet, where it occurs as either wet or dry gangrene – wet gangrene is foul-smelling, soft and red in appearance, while dry gangrene tends to be black (Figure 3.8). Finally, fibrinoid necrosis is a form of necrosis of smooth muscle; it is seen in vasculitis, particularly in beagle pain syndrome (Figure 3.9).

Characteristics of necrotic cells include karyorrhesis (fragmentation of the nucleus), karyolysis (dissolution of the nucleus) and pyknosis (nuclear shrinkage). Necrosis

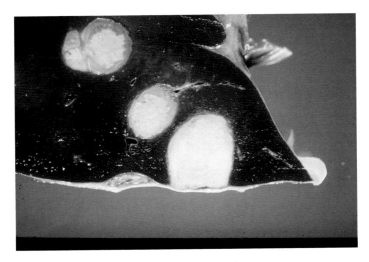

Figure 3.6 Caseous necrosis in the centres of ruminant liver abscesses.

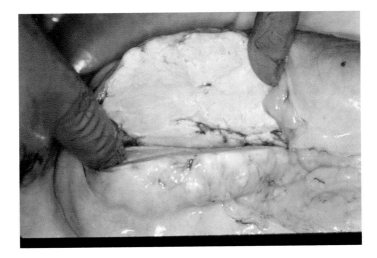

Figure 3.7 Fat necrosis in ruminant mesenteric fat tissue.

involves cell death, so ghost outlines of cells are seen under the light microscope. The body responds to necrosis by producing scars, erosions and ulcerations, and sometimes by abscess formation.

Mineralisation or calcification is the deposition of calcium salts in normal or necrotic tissues. It can be felt at necropsy, as the tissues will be hard and gritty, and it is generally visible as a white deposit. There are two types of calcification: dystrophic, which is local and occurs in dead and dying tissues (in the presence of normal serum calcium levels), and metastatic, which involves calcium deposition in normal tissues and usually reflects a disturbance of calcium metabolism or hypercalcaemia (i.e. an increase in serum calcium levels). Causes of hypercalcaemia include vitamin D toxicity, excess parathyroid hormone and kidney failure. Von Kossa and Alizarin red S special stains may be used to demonstrate mineralisation in tissues.

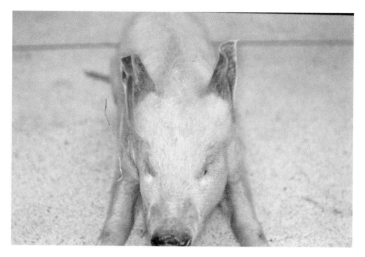

Figure 3.8 Dry gangrene in the black ear tips of a pig.

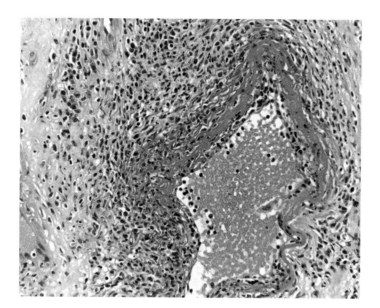

Figure 3.9 Fibrinoid necrosis in beagle pain syndrome, showing bright pink material in the blood vessel wall.

Apoptosis or programmed cell death (Figure 3.10) is a regulated cell suicide programme (Wyllie, 1997). It is an important process during organ development and provides physiological balance to mitosis. It can only be visualised under the light microscope. It occurs in normal tissue turnover (liver, pancreas), embryogenesis (the destruction of the webs between the digits of the human foetus) and endocrine-dependent tissue atrophy (e.g. prostatic atrophy after castration). Apoptosis is characterised by cell shrinkage, chromatin condensation, cytoplasmic blebs and, finally, phagocytosis of apoptotic bodies by adjacent macrophages. The TUNEL (terminal deoxynucleotidyl transferase

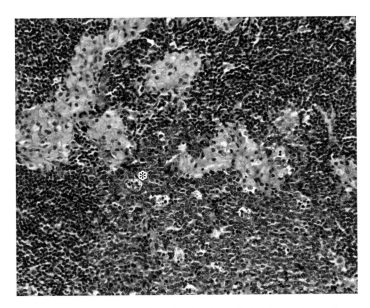

Figure 3.10 Apoptosis (*) in a mouse lymph node, showing cell shrinkage, chromatin condensation, cytoplasmic blebs and phagocytosis of apoptotic bodies by adjacent macrophages.

dUTP nick end labeling) method was developed to demonstrate breaks in DNA that occur in apoptosis. However, as these breaks may also occur during necrosis, the TUNEL technique does not distinguish between the two processes, and caspase immuno-histochemistry (IHC) is now thought to be more accurate than TUNEL in identifying apoptotic cells.

3.2 Inflammation

Study personnel will be familiar with inflammation, which can range from a mild serous nasal discharge seen in a beagle dog on an inhalation study to severe pneumonia and lung abscessation seen in a rat treated with an anticancer drug. Inflammation is the process that occurs when an animal attempts to remove an injurious agent and to repair the damaged tissue. Pathologists indicate inflammation in various organs and tissues by using the suffix '-itis' after the Greek word for the organ; thus, inflammation in the skin is called 'dermatitis' and inflammation in the kidney is called 'nephritis'. Inflammation is divided into acute and chronic, depending on the time frame. Acute inflammation is the immediate and early response to an injurious agent. The hallmark of acute inflammation is increased vascular permeability and a loss of fluid from the leaky blood vessels, whilst its causes include infection by bacteria, fungi, viruses and parasites, immune reactions, trauma, heat and cold, radiation, oxygen radicals, toxins, enzymes and chemicals (Brooks, 2010b).

The clinical signs of inflammation include red, hot, swollen tissues, pain and loss of function of an organ (Figure 3.11). The three major actions of the exudative phase of acute inflammation are an increase in blood flow, structural changes in the microvasculature that allow plasma proteins and leukocytes to leave the circulation and emigration

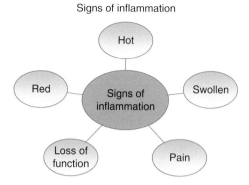

Figure 3.11 Clinical signs of inflammation.

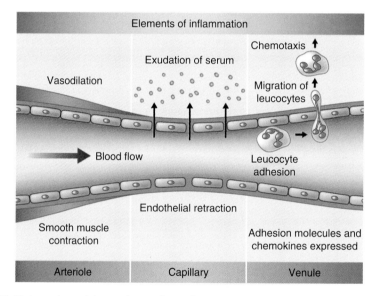

Figure 3.12 Major actions of the exudative phase of acute inflammation.

of the leukocytes (white blood cells, neutrophils) from the microcirculation into the area of injury (Figure 3.12). The main components of an inflammatory exudate include fibrin (a protein-rich secretion), serum and inflammatory cells (particularly neutrophils).

   The cells involved in inflammation are illustrated in Figure 3.13. The most important cells in acute inflammation are the polymorphonuclear leukocytes or neutrophils and macrophages (Kumar et al., 2010a). Neutrophils are granulated white blood cells with segmented, multilobed nuclei (which often have a horseshoe appearance); these move through the endothelium of the blood vessel to the area of inflammation, ingest (phago-cytosis) and kill microorganisms and release inflammatory mediators. Chemotaxis is the process whereby inflammatory cells move into an area of inflammation. Exogenous chemotactic stimuli include bacterial products, whilst endogenous stimuli include complement. Complement is a system of more than 20 proteins forming a cascade; each

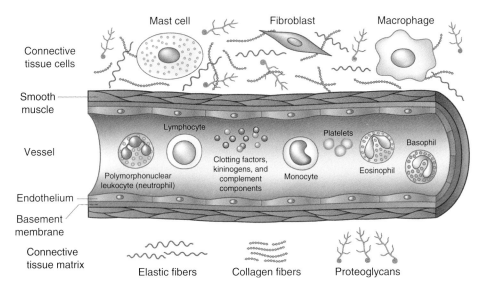

Figure 3.13 Cells involved in inflammation: neutrophils, macrophages, eosinophils, mast cells and basophils.

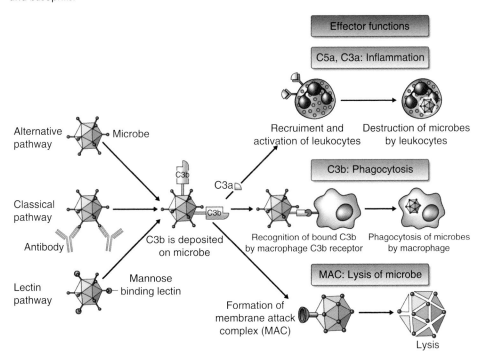

Figure 3.14 The complement system.

protein becomes activated (generally when antibody binds to antigen) and acts on the next inactive protein in order to convert it to an active form. Ultimately, complement ensures that microbes are converted to a state that makes phagocytosis by macrophages easier (Figure 3.14).

Leukocyte adhesion molecules, such as selections, allow neutrophils to move out of blood vessels. They are an important drug target (Luster et al., 2005) in the control of harmful inflammation. We know neutrophils are important in clearing inflammation because inherited defects in neutrophil adhesion and function cause recurrent bacterial infections (Ward et al., 2002). Chemical mediators of inflammation include histamine from mast cells, which causes increased vascular permeability, as well as serotonin, arachidonic acid metabolites, platelet-activating factor, the kinin system and the fibrinolytic system. In general, chemical mediators of inflammation originate from plasma or cells and execute their effects by binding to specific cell-surface receptors.

The different types of inflammation are serous (e.g. the fluid within a blister), fibrinous (MSB special stain is used to confirm the presence of fibrin) (Figure 3.15), purulent (Figure 3.1) and haemorrhagic. Fibrinous inflammation contains large amounts of the protein fibrin and is often seen on serous membranes, such as the pericardium (Figure 3.16), pleura and peritoneum. Purulent inflammation is associated with pus (i.e. necrotic cells, oedema fluid and neutrophils, which give the green colour to pus). Haemorrhagic inflammation contains large numbers of red blood cells.

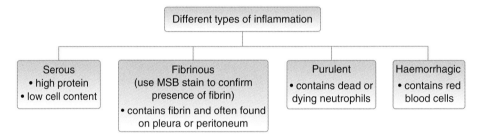

Figure 3.15 Types of inflammation.

Figure 3.16 Fibrinous inflammation around a ruminant heart.

Figure 3.17 Formation of fibrous tissue (scar tissue) (*) in the scars of old infarcts in a dog kidney.

Chronic inflammation is inflammation of prolonged duration, measured in weeks and months (Kumar et al., 2010a). It does not produce an exudate, and generally involves the formation of fibrous tissue (scar tissue) (Figure 3.17). Chronic and acute inflammation (with infiltration of mononuclear cells), tissue destruction and repair can occur simultaneously (Figure 3.18); examples include silica toxicity, inflammatory bowel disease, chronic kidney failure and tuberculosis. Chronic inflammation is characterised by the persistence of macrophages, lymphocytes and plasma cells (lymphocytes which produce antibodies). Eosinophils are motile white blood cells with segmented nuclei and pink granules that are associated with parasite infestations and allergic reactions. Macrophages are large cells with vacuolated cytoplasm that are able to phagocytose and digest microorganisms, as well as cause ongoing tissue destruction (Gordon and Taylor, 2005). Macrophages may join together to form multinucleate cells (Figure 3.19), which are noted in conjunction with foreign-body reactions and tuberculosis infections.

Granulomas and pyogranulomas occur in granulomatous inflammation; they are an attempt by the body to wall off and contain an offending agent that is difficult to remove. Granulomas are characterised by a central core of necrotic or foreign material, a layer of neutrophils and macrophages surrounded by lymphocytes and a layer of fibrous connective tissue. The most common example of granulomatous inflammation is the tubercle observed in human and animal tuberculosis. The resolution of chronic inflammation involves the organisation of the fibrin, growth of new capillaries, an influx of macrophages and the proliferation of fibroblasts, causing fibrosis, sometimes with subsequent adhesions involving collagen and scarring. Ultimately, inflammation functions to dilute, localise, destroy or remove the injurious agent and to induce replacement of the necrotic tissue with fibrous tissue (Figure 3.20).

Amyloidosis is a dense pink protein substance deposited in various tissues that stains pink with the Congo red special stain. Amyloid deposition has a number of causes, including plasma cell tumours (multiple myeloma), chronic inflammation and Alzheimer's disease.

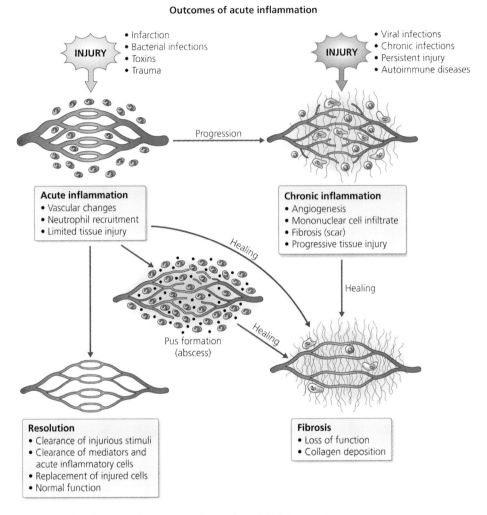

Figure 3.18 Simultaneous tissue destruction and repair in inflammation.

3.3 Circulatory Disturbances

Circulatory disturbances involve the blood. They are characterised by hyperaemia (an active process in which a greater amount of blood than is normal collects in an organ) and congestion (passive outflow of blood from the organ) (Figure 3.21). Haemorrhage is the escape of blood cells from a blood vessel due to vessel rupture; it can cause haematoma, if bleeding occurs into surrounding tissues, or epistaxis, if bleeding occurs from the nose. Petechiae are pinpoint haemorrhages (Figure 3.22), whilst ecchymosis are large bruises in tissue.

Oedema is characterised by increased fluid in the interstitial compartment. It causes soft swelling of tissues and pitting upon pressure, and is often seen in the limbs of animals. Inflammatory oedema occurs when there is a state of acute inflammation in

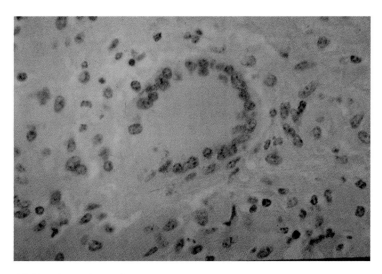

Figure 3.19 Multinucleate giant cell.

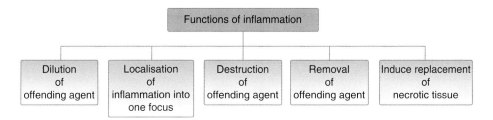

Figure 3.20 Functions of inflammation.

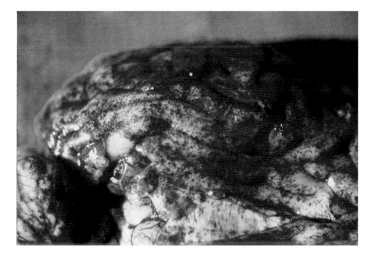

Figure 3.21 Hyperaemia on the surface of a ruminant brain.

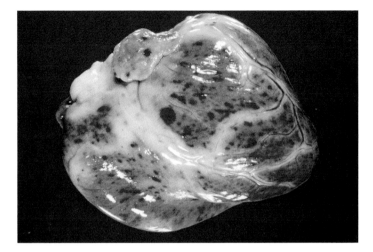

Figure 3.22 Petechiae on the surface of a canine heart.

an organ or tissue and consequently the fluid is high in protein. Noninflammatory oedema may be caused by increased blood pressure, decreased proteins in the blood (malnutrition or kidney failure) or lymphatic obstruction (due to the presence of tumours and scar tissue within the lymphatics). Oedema fluid can also accumulate in body cavities: hydrothorax is the accumulation of oedema fluid in the thoracic cavity (Figure 3.23), ascites is oedema of the abdominal cavity, anasarca is generalised oedema present below the skin and hydropercardium is the accumulation of oedema fluid within the pericardium.

Study personnel may see evidence of ischaemia (e.g. the dry gangrene causing a black tail with annular constrictions seen in ringtail in young mice). Ischaemia is hypoxia of

Figure 3.23 Hydrothorax is the accumulation of fluid in the thoracic cavity.

tissue due to reduction of the oxygenated blood supply, whilst infarction is the death of an area of tissue as a result of ischaemia. Ischaemia may be reversible, depending on its duration and the oxygen demands of the tissue. Causes of ischaemia include thrombosis, embolism, vascoconstriction, conmpression, vasculitis and blood vessel damage. Organs with an end blood supply (i.e. organs that are unable to develop a collateral blood supply, such as the brain, spleen, kidney and heart) are most susceptible to infarction, whilst those with an alternative blood supply (e.g. lung, muscle) are less susceptible (Brooks, 2010a). Ischaemia may be focal (e.g. a small wedge-shaped area in the kidney supplied by one arteriole) or global (e.g. gastric torsion due to twisting of the entire arterial supply).

Large blood clots are visible at necropsy and may be pathological (adhered to the vessel surface) or post-mortem (nonadherent to the vessel surface). Clotting functions to maintain blood in a fluid state in normal blood vessels and to rapidly produce a local plug at the site of vessel injury. Clotting (haemostasis) involves the conversion of fibrinogen to fibrin via the intrinsic (produces tissue factor) and extrinsic (produces thrombin) pathways (Figure 3.24). Together with platelets, fibrin forms a blood clot

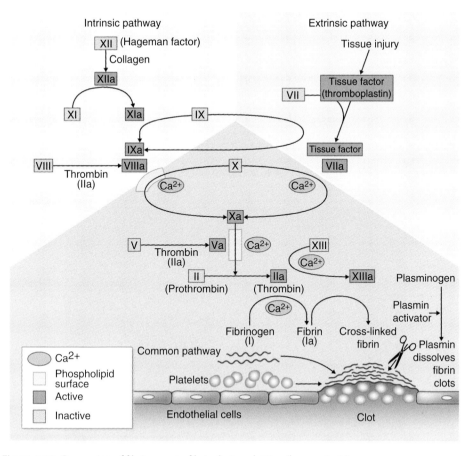

Figure 3.24 Conversion of fibrinogen to fibrin during clotting (haemostasis).

(Mackman, 2005). After the initial injury, arteriolar vasoconstriction occurs, and endothelial injury causes exposure of the subendothelial matrix, causing platelets to adhere to the vessel surface and tissue factor to activate the coagulation cascade. Platelets become activated and contract during clotting. Calcium and vitamin K (from the liver) are required at various steps in the coagulation cascade. Platelets (see Chapter 6) are small cytoplasmic structures with no nucleus that are present in the blood; they are formed from megakaryocytes in the bone. Disseminated intravascular coagulation (DIC) is a condition that may occur when toxins, severe burns, tumours and snake venoms cause widespread endothelial damage in an animal, leading to multiple small thromboses throughout the body. It results in the depletion of clotting factors and a predisposition to haemorrhage, with multiple haemorrhages developing throughout the body.

Thrombosis is a solid mass of fibrin formed by platelets in blood and is a pathological clotting reaction (Figure 3.25). Study personnel may see it in conjunction with indwelling catheters in continuous-infusion studies. Thrombi develop when there are changes to the three components of Virchow's triad: the vessel walls, blood flow and blood constituents, making the blood thicker (e.g. increased oestrogen). An embolus is defined as any detached intravascular solid, liquid or gaseous mass that is carried by the blood from its site of origin to a distant site. Emboli may consist of a piece of thrombus, small foci of tumour cells, parasites, fat, air, bone marrow, bacterial colonies enmeshed with inflammatory cells and fibrin, foreign material or keratin (from intravenous injections). A thrombus may experience propagation (the clot gets bigger), embolisation (small pieces of the clot break off and enter the bloodstream), dissolution/lysis (the clot dissolves), organisation (the clot undergoes fibrosis) or recanalisation (small canals develop within the clot, allowing partial blood flow).

Haemorrhage occurs when blood is lost from the circulatory system. It is easy to recognise in study animals. Causes of haemorrhage include trauma, chronic ulceration, rupture of a tumour or organ (e.g. liver, spleen) and coagulopathy (Brooks, 2010a). Coagulopathies are congenital or acquired conditions in which clotting fails to occur;

Figure 3.25 Thrombosis in a canine large blood vessel.

causes include vitamin K deficiency, liver disease (reduction in the production of clotting factors), warfarin, DIC and congenital diseases such as haemophilia A (deficiency of factor VIII). Different types of haemorrhage are defined macroscopically at necropsy. These include small pinpoint haemorrhages (petechiae), localised large areas of haemorrhage (haematoma), haemorrhage from the nose (epistaxis) (Figure 3.26), haemorrhage into the thoracic cavity (haemothorax) and haemorrhage into the abdominal cavity (haemoperitoneum). Melena is noted in the intestine where haemorrhage results in a black tarry faeces due to the digestion of blood products as they pass through the intestines to the rectum.

Shock is defined as a state of reduced blood pressure, and may present in different ways. Animals in shock may have pale, white mucous membranes, shallow breathing and collapse. Causes include loss of blood or fluid (hypovolemic shock), dilatation of blood vessels (vasodilatory shock) and DIC or vasodilatation due to toxins (septic shock).

Figure 3.26 Epistaxis in a dog.

3.4   Disorders of Tissue Growth

Study personnel will encounter enlarged livers at necropsy, generally caused by compounds such as peroxisome proliferator-activated receptors (PPARs), which produce large, swollen livers that bulge on cut surface. Growth responses by tissues and cells include hyperplasia, hypertrophy and atrophy (Figure 3.27). These findings occur in response to excessive physiologic stresses or pathologic stimuli, or to cellular adaptations aimed at preserving viability, modulating function and escaping injury.

Enlarged tissues (e.g. lactating mammary gland) occur as a result of general hyperplasia (i.e. an increase in number of cells), such as androgen-induced benign prostatic hyperplasia (Kumar et al., 2010b). Hyperplasia may have physiologic or pathologic causes and may lead to an increase in organ volume. It occurs only in cells that are capable of mitosis (i.e. dividing) and thus is not seen in nondividing/permanent cells, including most neurons and cardiac and skeletal muscle cells.

Hypertrophy is defined as an increase in the size of cells or an organ. It may be non-pathological (e.g. uterus in pregnancy, increased heart size in an athlete) or pathological (e.g. liver hypertrophy due to compounds such as PPAR agonists) (Lindblom et al., 2012). In hepatic hypertrophy, the pathologist sees large liver cells arranged around the portal tracts under the microscope, with large nuclei and abundant amounts of pink cytoplasm.

Atrophy is a decrease in size and function within an organ. A decrease in cell size is caused by loss of cell substance, which may be physiologic (e.g. the reduction in uterus size after birth) or pathologic (e.g. muscle atrophy when a broken limb is placed in a plaster cast). Loss of innervation (denervation atrophy) and pressure atrophy (e.g. tumour pressure) are also common examples of atrophy. Disuse of the tissue due to lack of nerve conduction or immobilisation of the limb results in a gradual reduction in the muscle tissue. Metaplasia is the transformation of one tissue type into another, such as the change from respiratory epithelium to squamous epithelium that occurs in the respiratory tract as a response to noxious stimuli such as smoking. These are all considered to be reversible changes in growth. Pathologists must examine tissues under the light microscope in order to determine whether an enlarged organ contains an increased

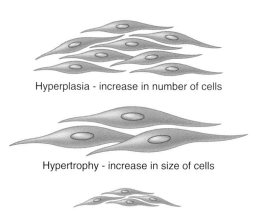

Hyperplasia - increase in number of cells

Hypertrophy - increase in size of cells

Atrophy - decrease in size and function of cells

Figure 3.27 Cell growth responses.

number of cells (hyperplasia) or whether the cells are larger (hypertrophy), as well as whether epithelial cell types have changed into a more primitive form (metaplasia).

3.5 Tissue Repair and Healing

It is easy to recognise healing, and most study personnel will be familiar with the presence of a thickened whitish scar forming at the edge of an ulcerative lesion in animal skin (especially in mice with bite wounds) (Figure 3.28). Damage to tissue results in a healing process that is divided between regeneration and repair (Kumar et al., 2010b). Regeneration leads to the complete restoration of the lost or damaged tissue and involves the proliferation of cells, as seen when erosion occurs in the small intestine. Tissues such as the skin and the gastrointestinal system heal continuously, provided stem cells are available (Mimeault et al., 2007) and the cellular matrix has not been damaged (Kumar et al., 2010b). Growth factors, angiogenesis and cytokines are all involved in tissue regeneration; the most important angiogenesis growth factor is vascular endothelial growth factor (VEGF) (Adams and Alitalo, 2007). Some tissues have a low level of regeneration and proliferate in response to injury (e.g. lymphocytes, kidney). In such cases, the regenerating cells are often basophilic (blue) when examined under the microscope. The blue colour tells us there is a lot of RNA in the cells, and thus that they are growing rapidly.

If tissue injury is severe and results in damage of the cells and the extracellular network, healing via regeneration will not take place. Cells that are unable to regenerate are unable to divide; these are termed 'terminally differentiated cells' and include most

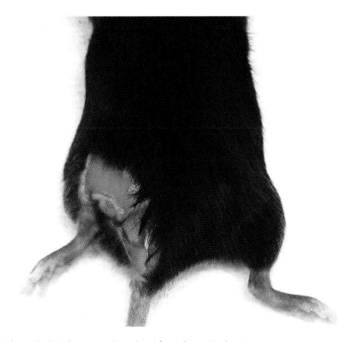

Figure 3.28 Thickened whitish scar at the edge of an ulcerative lesion.

neurons in the brain and muscle cells in the heart. The heart heals by scar formation or fibrosis (the extensive deposition of collagen and extracellular matrix components in areas of necrosis), followed by scar proliferation (fibroblasts and capillaries) and maturation. Healing is influenced adversely by poor nutrition, diabetes mellitus, inadequate blood supply, movement and corticosteroids. Generally, most lesions involving fibrosis and scar tissue can be considered chronic, whilst those involving congested blood vessels and exudation of fluid are likely to be acute.

3.6 Neoplasia

Personnel involved in carcinogenicity studies will be very familiar with tumours of all sizes, occurring at many different sites. Tumours are seen in rats and mice after approximately 1 year, including the large fibroadenomas (benign mammary tumours that often grow to a large size) found in rats. Neoplasia is the proliferation or new growth of abnormal cells to form a cancer or tumour (neoplasm); neoplastic cells grow and divide in an uncontrolled manner. Tumours are either benign or malignant (Table 3.1): benign tumours generally do not spread, are encapsulated and can be removed by surgery, whilst malignant tumours destroy tissue surrounding their original site (invasive), can spread to other organs in the body (metastasis) and display variation in the size of their cells (pleomorphism) which have abnormal nuclear morphology.

Tumours are named according to the tissue from which they arise. Benign tumours are further modified by the use of the suffix '-oma' after the tissue name, whilst malignant tumours are modified by the suffix '-carcinoma' or '-sarcoma': 'carcinoma' indicates a malignant tumour of epithelial cells, whilst '-sarcoma' indicates a malignant tumour of mesenchymal cells. The term 'adeno-' indicates that the tumour is derived from glandular tissue; benign epithelial tumours are called 'adenomas', whilst malignant glandular tumours are called 'adenocarcinomas'. Perplexingly, lymphoma, melanoma, mesothelioma and seminoma are not always benign and are in fact generally malignant tumours! Tumours demonstrating a mixture of mesenchymal and epithelial elements are often called 'mixed tumours' or 'complex adenomas'.

Cells are able to divide normally using a process of mitosis and the cell cycle, which is controlled by growth factors and growth inhibitors. Benign tumours generally have a

Table 3.1 Differences between benign and malignant tumours.

	Benign	Malignant
Behaviour	Expansile growth only	Expansile and invasive growth
	Grows locally	May metastasise
Histology	Resembles cell of origin (well differentiated)	May show loss of cellular differentiation (anaplasia)
	Few mitoses	Many mitoses, some of which are abnormal and bizarre
	Normal or slight increase in ratio of nucleus to cytoplasm	High nucleus to cytoplasm ratio
	Cells are uniform throughout the tumour	Cells vary in shape and size (cellular pleomorphism) or nuclei vary in shape and size (nuclear pleomorphism)

low mitotic index, whilst malignant tumours generally have a high one. Malignant tumours undergo local and distant metastasis (spread via either blood or lymph and serosal surfaces). Malignant tumours often metastasise to local draining lymph nodes. Tumours that metastasise via the blood tend to lodge in the lung and liver, because caval and portal veins carry blood to the lungs and liver, respectively. It is thought that cancers are maintained by cancer stem cells, which are resistant to chemotherapy and radiotherapy because of their low rate of division (Jordan et al., 2006).

The genes that are generally mutated in tumours include growth-promoting, growth-inhibiting, apoptosis and DNA-repair genes. Tumour cells lose growth-inhibiting genes or experience activation of growth factors, leading to uncontrolled growth. In addition, tumours are able to evade apoptosis, can replicate limitlessly, have excellent angiogenesis properties and have defects in DNA repair. Hundreds of chemicals are carcinogenic, including soot, cyclophosphamide (ironically an anticancer drug!), benzidine and aflatoxin B1, in addition to radiation and ultraviolet (UV) light (the sun is thought to induce melanoma, squamous cell carcinoma and basal cell carcinoma in fair-skinned Europeans living near the equator), alcohol, high-fat diet, asbestos, tobacco, viruses, older age, chronic inflammation, immunosuppression and genetic predisposition. Papilloma viruses and *Helicobacter pylori* bacteria cause tumours in humans, whilst Marek's disease virus causes lymphoma in chickens.

Angiogenesis factors encourage the growth of blood vessels in order to service a tumour, and thus angiogenesis inhibitors can restrict the growth of tumours by restricting their blood supply. Small tumour masses must be able to invade the extracellular matrix, enter the bloodstream, adhere to the basement membrane of the blood vessel, grow out of the blood vessel and develop into a new tumour focus with its own blood supply in a distant tissue. Recently, there has been interest in miRNAs: small, noncoding, single-stranded RNAs about 22 nucleotides in length that are incorporated into the RNA-induced-silencing complex (Stricker and Kumar, 2010). Drugs that can inhibit or augment the functions of miRNAs are potentially useful as chemotherapeutic agents, and pharmaceutical companies are currently undertaking research in this area.

The formation of malignant tumours requires the mutational loss of many genes (not just one), including those that regulate apoptosis (Chichowski and Hahn, 2008). Tumours can evade immune surveillance in the body through a loss or reduction of major histocompatibility complex (MHC) molecules (MHC-class 1 proteins on the surfaces of cells), a lack of costimulation and masking of tumour antigens (Stricker and Kumar, 2010). Pathologists must examine tumours under the light microscope to establish whether they are benign or malignant.

3.7 Immune System

The body has complex defence mechanisms – generally, immune responses – which protect against numerous insults, including bacteria, viruses, fungal spores and foreign material (which enters via the mouth, lungs, urogenital tract or breaks in the skin).

Study personnel will be familiar with lymphocyte cell counts in the blood and reductions in the size of lymphoid organs such as the spleen and thymus, which indicate immunodeficiency caused by treatment. The innate immune system is made up of chemical mediators of inflammation, plasma proteins, phagocytic inflammatory cells and natural killer cells (large granular lymphocytes). The innate immune system does not recognise specific bacteria or viruses, but can respond generally to infectious agents; the adaptive immune response (acquired immunity), on the other hand, responds to specific infectious agents and may develop memory cells to a particular agent.

The adaptive immune system consists of humoural immunity, where cells (lymphocytes) produce antibodies and cell-mediated immunity. Lymphocytes are associated with the lymphoid organs of the body. They are divided into B- (originate in the bone marrow) and T- (originate in the thymus) cells, and T-cells are further divided into CD4+ (helper cells) and CD8+ (cytotoxic cells) T-cells. MHC class I and II restriction ensures that T-cells only recognise other cells with the antigen attached to them (Klein and Sato, 2000a,b). Dendritic cells (CD1+) are involved in antigen presentation and possess fine cytoplasmic processes. Plasma cells develop from antigen-stimulated B-cells and produce antibodies, which are classified into isotypes according to their heavy-chain characteristics.

The binding of antibody to antigen will activate the complement cascade. Complement is a system of proteins that form a cascade, whereby each protein becomes activated (generally when antibody binds to antigen) (Figure 3.14) and then acts on the next inactive protein in order to convert it to an active form. Ultimately, complement ensures that microbes are converted to a state that makes phagocytosis by macrophages easier. The T-cells have specific antigen receptors on their surfaces and are involved in memory, in presenting the antigen to antigen-presenting cells and in cytotoxic cell responses (destruction of infected cells). Cytokines are the messengers of the immune system, and induce and regulate immune responses. CD4 helper cells produce TH1, TH2 and TH17 subsets of lymphocytes, which differ in the cytokines they produce (TH1 produces interferon gamma, TH2 produces interleukin 4 and TH17 produces interleukin 17) (Reiner, 2007).

Abnormalities in the immune system consist of primary immunodeficiencies, which are congenital and include abnormalities in immunoglobulins, immune cells and complement. Severe combined immunodeficiency (SCID) makes mice highly susceptible to infection, since they do not possess T- or B-lymphocytes. Secondary immunodeficiency is caused by physiological states (e.g. stress, pregnancy), chronic disease, viruses (e.g. simian immunodeficiency virus), tumours and drugs (e.g. corticosteroids). Immunodeficiency or immunosuppression makes an animal more susceptible to infectious agents that are normally resisted. Immunodeficient animals are particularly susceptible to pneumonia caused by *Pneumocystis* species, as well as cytomegalovirus.

Hypersensitivity is an exaggerated immune response to stimuli that would not usually be problematic, including reactions to insect bites. There are different types of hypersensitivity:

- Type I hypersensitivity involves the immunoglobulin-activated degranulation of mast cells (produces histamine) and the presence of eosinophils, and includes anaphylaxis (shock, oedema and difficulty breathing), insect bite reactions and hay fever.
- Type II hypersensitivity involves the inappropriate killing of the body's cells (e.g. blood transfusion reactions, where erythrocytes are destroyed, causing severe anaemia). Various drugs can cause haemolytic anaemia by attaching to the surface membrane of red blood cells and causing antibodies to form against the drug–membrane complex. Study personnel will encounter compound-mediated haemolysis – resulting

Interelationship between different systems in inflammation

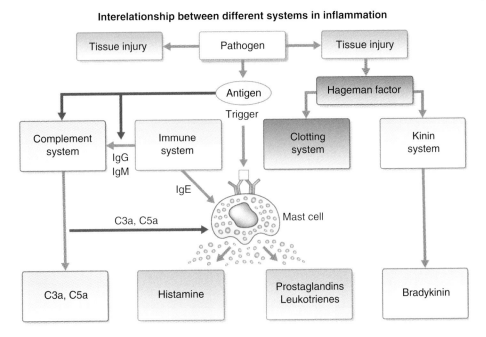

Figure 3.29 Interrelationship between different systems in inflammation.

in anaemia (pale white mucous membranes), dark urine (haemoglobin from broken-down red blood cells) and a large spleen (splenomegaly, caused by the spleen breaking down the erythrocytes and attempting to produce new immature red blood cells at the same time) – at necropsy.

- Type III hypersensitivity involves damage by antigen/antibody/complement complexes. These form immune complexes, which may lodge in the blood vessels of the glomerulus of the kidney, stimulating inflammation and leading to glomerulonephritis.
- Type IV (delayed-type)_hypersensitivity takes several days to weeks to develop and involves lymphocytes (predominantly T-cells) and macrophages, generally resulting in granuloma formation. The tubercle formed in tuberculosis infections caused by *Mycobacterium* species is an example of a type IV response, and is characterised by a central core of caseating material surrounded by neutrophils, which are in turn surrounded by macrophages and fibroblasts.

Figure 3.29 illustrates how the immune system fits into the other components of inflammation.

References

Adams, R.H. and Alitalo, K. (2007) Molecular regulation of angiogenesis and lymphangiogenesis. *National Reviews in Molecular and Cellular Biology*, 8(6), 464–78.

Brooks, H. (2010a) Circulatory disorders. In: Brooks, H. (ed.). *General Pathology for Veterinary Nurses*, Wiley-Blackwell, Hoboken, NJ.

Brooks, H. (2010b) Inflammation. In: Brooks, H. (ed.). *General Pathology for Veterinary Nurses*, Wiley-Blackwell, Hoboken, NJ.

Chichowski, K. and Hahn, W.C. (2008) Unexpected pieces to the senescence puzzle. *Cell*, 133(6), 958–61.

Golstein, P. and Kroemer, G. (2007) Cell death by necrosis: towards a molecular definition. *Trends in Biochemical Science*, 32(1), 37–43.

Gordon, S. and Taylor, P.R. (2005) Monocyte and macrophage heterogeneity. *National Reviews in Immunology*, 5, 953–64.

Jordan, C.T., Guzman, M.L. and Noble, M. (2006) Cancer stem cells. *New England Journal of Medicine*, 355(12), 1253–61.

Klein, J. and Sato, A. (2000a) The HLA system. First of two parts. *New England Journal of Medicine*, 343(10), 702–9.

Klein, J. and Sato, A. (2000b) The HLA system. Second of two parts. *New England Journal of Medicine*, 343(11), 782–6.

Kumar, V., Abbas, A.K., Fausto, N. and Aster, J.C. (2010a) Cellular responses to stress and toxic insults: adaption, injury and death. In: Kumar, V. (ed.). *Robbins & Cotran Pathologic Basis of Disease*, 8th edn, Saunders Elsevier, Philadelphia, pp. 3–42.

Kumar, V., Abbas, A.K., Fausto, N. and Aster, J.C. (2010b) Tissue renewal, repair and regeneration. In: Kumar, V. (ed.). *Robbins & Cotran Pathologic Basis of Disease*, 8th edn, Saunders Elsevier, Philadelphia, pp. 79–110.

Lindblom, P., Berg, A.L., Zhang, H., Westerberg, R., Tugwood, J., Lundgren, H., Marcusson-Ståhl, M., Sjögren, N., Blomgren, B., Öhman, P., Skånberg, I., Evans, J. and Hellmold, H. (2012) Tesaglitazar, a dual PPAR-α/γ agonist, hamster carcinogenicity, investigative animal and clinical studies. *Toxicologic Pathology*, 40(1), 18–32.

Luster, A.D., Alon, R. and von Andrian, U.H. (2005) Immune cell migration in inflammation: present and future therapeutic targets. *Nature Immunology*, 6(12), 1182–90.

Mackman, N. (2005) Tissue-specific hemostasis in mice. *Arteriosclerosis, Thrombosis, and Vascular Biology*, 25, 2273–81.

Mimeault, M., Hauke, R. and Batra, S.K. (2007) Stem cells: a revolution in therapeutics-recent advances in stem cell biology and their therapeutic applications in regenerative medicine and cancer therapies. *Clinical Pharmacology and Therapeutics*, 82(3), 252–64.

Reiner, S.L. (2007) Development in motion: helper T cells at work. *Cell*, 129(1), 33–6.

Stricker, T.P. and Kumar, V. 2010. Neoplasia. In: Kumar, V. (ed.). *Robbins & Cotran Pathologic Basis of Disease*, 8th edn, Saunders Elsevier, Philadelphia, pp. 3–42.

Ward, D.M., Shiflett, S.L. and Kaplan, J. (2002) Chediak-Higashi syndrome: a clinical and molecular view of a rare lysosomal storage disorder. *Current Molecular Medicine*, 2(5), 469–77.

Wyllie, A.H. (1997) Apoptosis: an overview. *British Medical Bulletin*, 53(3), 451–65.

4

Common Spontaneous and Background Lesions in Laboratory Animals

Elizabeth McInnes

Cerberus Sciences, Thebarton, SA, Australia

Learning Objectives

- Understand the different causes of nonproliferative background lesions in laboratory animals.
- Identify specific background lesions in mice, rats, dogs, non-human primates, rabbits and minipigs.
- Identify spontaneous causes of death in rats and mice.
- Discover whether all significant findings are test article-related.

Tyro study personnel will be rapidly introduced to the presence of background lesions in laboratory animals used for safety studies and will be confronted by the fact that sometimes these lesions are ignored, sometimes they are exacerbated at the end of a study (and thus become test article-related findings) and sometimes they are unique to a particular animal or procedure. This chapter will attempt to define and discuss the causes of background lesions, before looking at a few changes that are unique to mice, rats, beagle dogs, non-human primates, rabbits and minipigs. Common causes of death or euthanasia of older rodents on carcinogenicity studies will be discussed, as will the significance of test article-related background findings. Clear histopathological and macroscopic photographs illustrating some of these changes will be provided, although histopathology is not meant to be a skill for study personnel. The changes described here are those seen most commonly; further detailed information is provided in Johnson et al. (2013).

Background (incidental or spontaneous) lesions are pathology findings that are usually thought of as a change in tissue morphology outside of the range of normal variation for a particular species or strain (Long and Hardisty, 2012). Background changes can be congenital or hereditary; normal variations of findings that are unique to the histology of an animal species; related to trauma or normal ageing; or physiologic or hormonal changes (McInnes, 2012a; McInnes and Scudamore, 2014).

Congenital lesions are present at birth and probably represent abnormalities in normal embryogenesis and organ migration of the unborn animal. Squamous cysts – often

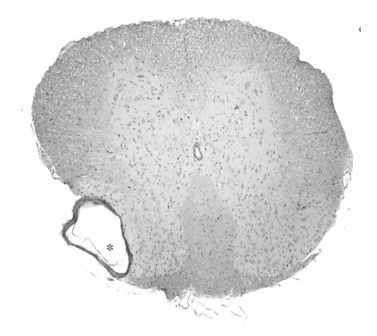

Figure 4.1 Squamous cyst (*) in a mouse spinal cord.

containing keratin, since they are lined with the same epithelium found on the skin surface – are common congenital background lesions in the stomach and central nervous system (Figure 4.1). Minor developmental anomalies can also include ectopic tissue (tissue found in unexpected places; Figure 4.2).

Background lesions also include normal ageing changes and degenerative conditions. Ageing background changes include chronic nephropathy (chronic kidney disease in rats and mice), cardiomyopathy (chronic heart disease in rats and mice) and polyarteritis (chronic inflammation of blood vessels in mice) (McInnes and Scudamore, 2014). All of these findings can be exacerbated by certain treatments, so a knowledge of what is normal for the strain and age of rodent being evaluated is crucial to interpretation of these changes. Study personnel should also be aware that age lesions can include younger animals and that the thymus is always bigger in younger animals as it has not yet undergone regression.

Background lesions influenced by ageing also include spontaneous tumours (not induced by external treatments or compounds) in older laboratory animals. It is helpful for study personnel to know about the incidence of these naturally occurring tumours in control animals, in order to assess whether an external treatment has caused an increased incidence of a particular tumour type (Hardisty, 1985). There are extensive references on the expected background incidence of spontaneous tumours and their descriptions in the various strains of rodents; many of these are summarised on the goRENI website (www.goreni.org).

Some background lesions are caused by infectious agents. For instance, lesions consisting of alveolar macrophages and lung wall thickening are reported in the lungs of immunocompetent young Han Wistar rats and have recently been shown to be

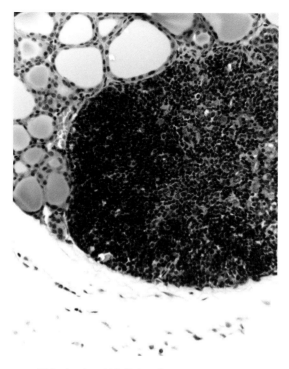

Figure 4.2 Bluish thymus within the thyroid follicles of a mouse.

associated with the presence of *Pneumocystis carinii* DNA (Livingston et al., 2011; Henderson et al., 2012). These lesions may mask or confuse research findings, particularly in inhalation studies.

Background findings may be traumatic in origin and can include fractures, bite wounds and foot lesions. Gavage injury is not a background change as such, but is a commonly encountered incidental, non-drug-related finding in rats and mice. It can consist of damage to the oesophagus and surrounding pleural tissues due to the penetration of the gavage tube through the wall of the oesophagus (Bertram et al., 1996).

Artefacts are a subgroup of background changes. They are important because they may occur as a result of faulty collection and processing of tissue specimens (McInnes, 2012b). Knowledgeable study personnel can recognise these artefacts in reports (such as air bubbles trapped below the cover slip on a glass slide) and take steps to eliminate them, as artefacts are produced at different stages of the harvesting and processing of tissue samples.

Background lesions are also associated with physiological variability, including reproductive senescence and sexual maturity, since significant variability in the appearance of the reproductive organs occurs as a result of normal reproductive cyclicity. Examples include changes associated with oestrus, sexual maturation and reproductive ageing in both sexes.

Normal physiology in rodents may also result in background changes such as haematopoiesis (the presence of small clusters of immature red blood cells) in the spleen and liver (Figure 4.3), failure of growth plate closure in the long bones, continuous eruption

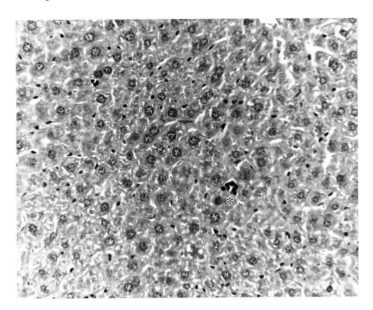

Figure 4.3 Haematopoiesis (*) in the liver.

of incisor teeth or thymic involution. Stress is a normal physiological response, and stress during animal studies may be caused by noise, changes in temperature, handling, dosing, restraint, sample collections, transport or group housing of animals (Everds et al., 2013). It can lead to reduced total body weight or body weight gain, reduced food consumption, changes in organ weights (e.g. decreased thymus and spleen weight, increased adrenal weight), lymphocyte depletion in the thymus and spleen, increased or decreased white blood cell levels in the blood, gastric ulceration and reproductive organ atrophy; Everds et al., 2013).

4.1 Rats

Rats display similar background lesions to mice, including mineralisation of the kidney papilla (the presence of small foci of calcium in the medullary papilla) and inflammatory cell foci in the liver. Rats have a specialised sebaceous gland situated at the base of the external ear canal called the Zymbal's gland, as well as Harderian glands behind the eye, for which there is no human equivalent. The rodent stomach subdivides into the glandular and nonglandular stomach; lesions in the nonglandular part may not have relevance to humans, who only possess a glandular stomach. In addition, the preputial or clitoral glands situated between the penis and the rectum in the mouse and rat have no human equivalent and may not be relevant to human patients.

Chronic progressive nephropathy (CPN) is a common spontaneous kidney disease of ageing rats, including Fischer 344 rats (Dixon et al., 1995) and mice. It is more common in male rats, and is exacerbated by a high-protein diet (Montgomery and Seely, 1990). CPN is also directly influenced by total food consumption over a lifetime (Keenan et al., 2000). Administration of certain drugs may exacerbate or hasten the lesions of CPN, particularly in long-term studies. CPN is observed commonly in ageing mice, and is

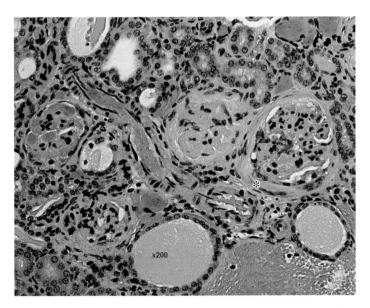

Figure 4.4 Thickening of the glomerular basement membranes and periglomerular fibrosis (*) in an ageing mouse, indicative of CPN.

characterised by thickening of the glomerular basement membranes and periglomerular fibrosis (Percy and Barthold, 2007) (Figure 4.4). The age of onset is later in mice than in rats, and the incidence is not as high (Frazier, 2013).

The presence of alveolar macrophages in the lung alveoli is a background change in rats and mice, and thus interpretation of a drug-induced alveolar macrophage increase may be difficult. Spontaneous alveolar macrophage aggregates are generally observed randomly dispersed throughout the lung tissue, whereas induced macrophage responses tend to be located at the bronchoalveolar junction (Lewis and McKevitt, 2013).

Cardiomyopathy (chronic heart disease) is a common lesion of older rats. It is found most frequently in the myocardium of the left ventricular wall, and is more common in male rats (MacKenzie and Alison, 1990). The cause of the disease is unknown, but severity and age of onset are influenced by diet, environment and stress (MacKenzie and Alison, 1990). The lesion is characterised initially by necrotic cardiomyocytes, surrounded by inflammatory cells. Later, fibrosis or scarring is prominent in ageing rats, generally in the left ventricle, intraventricular septum and papillary muscle.

4.2 Mice

Common background lesions in mice include the sexual dimorphism of the salivary gland (which produces prominent pink granules in the submandibular salivary gland of the male animals), adenomatous hyperplasia of the glandular stomach (which produces a thickened mucosa filled with dilated gastric glands; Figure 4.5) and enlarged nuclei in the cells of the liver (Figure 4.6). The spleen is often enlarged at necropsy (splenomegaly), and there are a variety of reasons for this lesion (Figure 4.7). Two of the most common are extramedullary haematopoiesis (EMH; foci of immature nucleated red blood

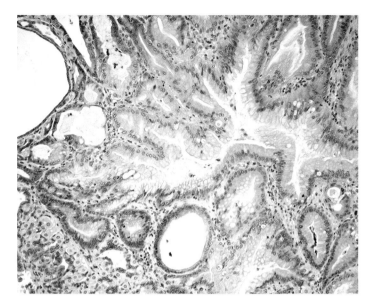

Figure 4.5 Thickened mucosa filled with dilated gastric glands, indicating adenomatous hyperplasia of the glandular stomach.

cells) and extramedullary granulopoiesis (foci of immature granulocytes.i.e. neutrophils). EMH is a normal finding in the red pulp of the spleen in mice, and is usually more prevalent in young animals and females (Suttie, 2006). When it is severe, the cause is increased demand as a result of anaemia (due to chronic bleeding, cancer or inflammation), with increased granulopoiesis observed in cases of abscess formation. Lymphoma

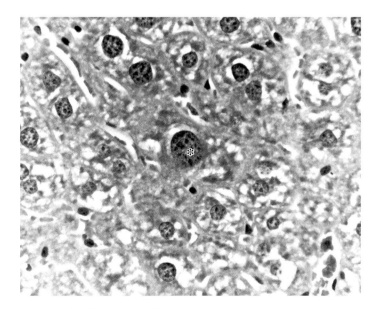

Figure 4.6 Enlarged nuclei (*) in the cells of the liver.

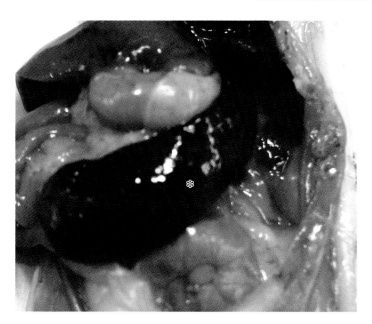

Figure 4.7 Enlarged mouse spleen (*) at necropsy.

is a common tumour in mice and also causes splenic enlargement, but it generally also involves the thymus, liver and lymph nodes.

Some strains of mice, particularly from a C57Bl/6 background, are prone to developing loss of hair and ulcerative skin lesions, which often start in the dorsal neck region (Figure 4.8) (Sundberg et al., 2011). Traumatic lesions may occur as a result of accidental damage or fighting, particularly when some strains are group-housed. These lesions may be confined to superficial skin damage (often to appendages) or may progress to abscesses (often in the head region).

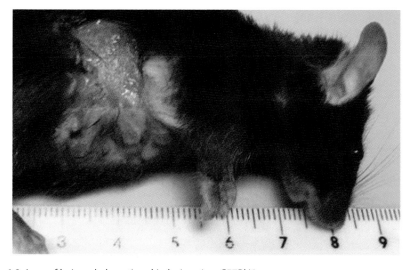

Figure 4.8 Loss of hair and ulcerative skin lesions in a C57Bl/6 mouse.

4.3 Dogs

Dogs are popular for cardiovascular and ocular studies. Dogs and other vertebrates have a tapetum lucidum: a layer of tissue in the eye that reflects visible light back through the retina, thus improving night vision. Study personnel should note that non-human primates (and humans) do not have this structure. One of the most important and dramatic background lesions seen in beagle dogs is beagle pain syndrome, which is characterised by inflammation of the arteries (arteritis). Arteritis is found sporadically in about 5–10% of young beagles (Grant Maxie and Robinson, 2007), occurring more commonly in males, and affects either single or multiple vessels, most commonly in the epididymides, thymus and the coronary arteries of the heart (Son, 2004a). In the majority of cases in young dogs, the lesion does not appear to have a cause (Hartman, 1987) and occurs in the absence of any clinical signs.

The beagle has less efficient and more irregular spermatogenesis than rodents, resulting in atrophic tubules with no germ cells (tubular hypoplasia) and hypospematogenesis (one or more generations of germ cell are missing) in 30% of normal beagle dogs aged between 6 and 36 months (Rehm, 2000). This is a problem, as it is easy to confuse these spontaneous findings with those caused by administered compounds. Multinucleate degenerating germ cells are also seen in the normal dog testis. Sperm granuloma are caused by initial sperm stasis, with later rupture upon the exposure of sperm to the body's inflammatory system. As sperm contains half the number of chromosomes found in normal tissues, it is perceived as foreign, and a foreign-body reaction and granuloma result. All of these changes are only noted under the light microscope and will not be visible at necropsy.

4.4 Minipigs

Minipigs possess skin that is structurally similar to that of humans and have a similar composition of the stratum corneum. They are thus often used for skin-toxicity studies (Swindle et al., 2012). Minipigs have certain anatomical differences from other animals that can prove problematic, however. The thyroids on the ventral surface of trachea at the thoracic inlet are hard to locate. They are often damaged in this position by venipuncture procedures, which can affect thyroid hormone levels. In addition, the parathyroids in the minipig are not attached to thyroids, but are located close to the carotid artery bifurcation in the thymus, fat or connective tissue, often just caudal to the hyoid bone (Jeppesen and Skydsgaard, 2015). The minipig parathyroids have a slightly reddish colour, and the texture of the parathyroid is different to that of the thymus. Furthermore, the thymus is located in the cervical region (i.e. in the neck and cranial thorax, adjacent to the trachea). The minipig lymph nodes are reversed compared with other animals, with a medulla on the outside of the node and the cortex in the centre of the node.

Necrosis of the gall bladder (cholecystitis) is an unusual background lesion found in minipigs (McInnes, 2012c). It is generally observed at necropsy, where the necrotic gall bladder appears thickened. Histopathologically, the lesion is characterised by necrosis of the mucosa, with large numbers of inflammatory cells extending into the submucosa and smooth-muscle layers. Cholecystitis is not associated with clinical signs or changes in clinical pathology parameters.

4.5   Non-Human Primates

Common background lesions in non-human primates include pigmented macrophages in the small intestine and accessory splenic tissue in the pancreas (Chamanza, 2012). Yellowish-brown to dark-brown pigment in lung macrophages is often found distributed around the blood vessels or bronchioles of the lungs of macaques. Although iron-positive pigment associated with the lung mite *Pneumonyssus simincola* is occasionally encountered, especially in wild-caught macaques, the majority of the brown pigment present in the lungs of purpose-bred macaques is usually not associated with any parasite sections or pulmonary parenchymal pathological changes, and is therefore believed to be anthracosis. Anthracosis, or pneumoconiosis, is caused by the inhalation of atmospheric carbon particles by laboratory non-human primates housed near or within urban areas. Similar pigment can be encountered within macrophages in the bronchial or mediastinal nodes and should be differentiated from tattoo pigment.

4.6   Rabbits

Common background lesions in rabbits include squamous (keratin) proliferation in the prostate and the interstitial gland in the rabbit ovary (Mori and Matsumoto, 1973). Rabbits have an appendix made up of a large collection of lymphoid nodules at the ileocaecal valve, which is visible at necropsy.

4.7   Experimental Procedures

Intravenous studies often produce increased keratin and hair granulomas in the lung, as these substances are introduced from the skin into the blood by the injection procedure. Continuous-infusion studies result in background lesions such as chronic inflammation, fibrosis, foreign-body reactions to sutures (granuloma formation), abscess, thrombosis and haemorrhage at the site of the catheter insertion into the blood vessel (Weber et al., 2011). These findings should be distinguished from test article-related lesions. The bleeding of rodents using the retro-orbital sinus behind the eye or the tongue vein also produces areas of inflammation in the optic nerve and ventral tongue, which lead to characteristic background lesions.

4.8   Causes of Death in Rats and Mice

Animals that die during studies must have a cause of death or demise attributed to them by the pathologist through light microscopic examination of the tissues. Renal pathology is one of the more common causes of non-neoplastic death in older mice from many strains (McInnes and Scudamore, 2015). Obstructive uropathy is also common in male mice, occurring due to obstruction of the lower urinary tract, often as a result of fighting in multiple-housed male mice. Other causes of death, such as haemorrhagic ovarian cysts, cardiomyopathy, arteritis and liver necrosis, are reported

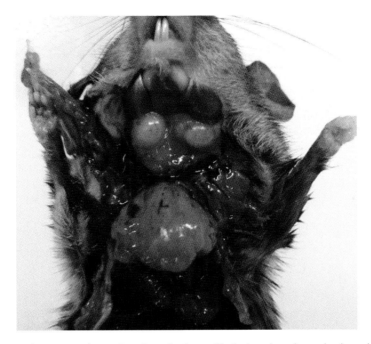

Figure 4.9 Lymphoma, manifesting in enlarged submandibular lymph nodes and enlarged thymus.

occasionally in mice in long-term studies (Son, 2003), whilst lesions such as ulceration, abscessation and pyogranulomatous inflammation of the skin or muscle make up a high proportion of causes (Son, 2003).

Although varying in incidence, the most common tumours seen in most strains of mice tend to be lymphoma (Figure 4.9), hepatocellular tumours, lung tumours, vascular tumours, pituitary tumours and Harderian gland tumours (McInnes and Scudamore, 2015). Son and Gopinath (2004) reported that between 50 and 80 weeks, the most common tumour was lymphoma, followed by lung adenoma. Some tumours may cause no significant clinical signs even though they reach a very large size (liver tumours), whilst some benign tumours may cause significant clinical signs that result in euthanasia of the animal (such as Harderian tumours behind the eye, which cause protrusion and subsequent infection of the eyeball).

CPN (chronic progressive nephropathy) is one of the most common causes of non-neoplastic death in older rats. A small number of animals will also die as a result of trauma or anaesthetic accident, or because of pododermatitis (Son, 2004b). In Han Wistar rats, the most common and earliest occurring tumour in both sexes is malignant lymphoma (Son et al., 2010), followed by brain tumour in male and mammary tumour in female animals. In Sprague Dawley rats, the most common early tumour is pituitary tumour in females (Figure 4.10) (Son et al., 2010).

4.9 Conclusion

This chapter has discussed the wide range of background lesions found in all types of laboratory animal used in toxicity studies. When background findings are significantly

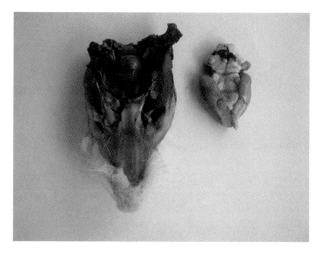

Figure 4.10 Pituitary tumour in a female Sprague Dawley rat.

increased in treated animals, study personnel should exercise great caution and work with the study pathologist in reporting these results. Although treatment may exacerbate a background finding (such as CPN), it is more likely that an increased incidence of a background finding in treated animals is fortuitous and not related to treatment.

References

Bertram, T.A., Markovits, J.E. and Juliana, M.M. (1996) Non-proliferative lesions of the alimentary canal in rats. GI-1. In: Guides for Toxicologic Pathology, STP/ARP/AFIP, Washington, DC.

Chamanza, R. (2012) Non-human primates. In: McInnes, E.F. (ed.). Background Lesions in Laboratory Animals, Saunders Elsevier, Edinburgh, pp. 1–15.

Dixon, D., Heider, K. and Elwell, M.R. (1995) Incidence of nonneoplastic lesions in historical control male and female Fischer-344 rats from 90-day toxicity studies. *Toxicologic Pathology*, 23, 338–48.

Everds, N.E., Snyder, P.W., Bailey, K.L., Bolon, B., Creasy, D.M., Foley, G.L., Rosol, T.J. and Sellers, T. (2013) Interpreting stress responses during routine toxicity studies: a review of the biology, impact, and assessment. *Toxicologic Pathology*, 41, 560–614.

Frazier, K. (2013) Urinary system. In: Sahota, P.S., Popp, J.A., Hardisty, J.F. and Gopinath, C. (eds). Toxicologic Pathology: Nonclinical Safety Assessment, CRC Press, Boca Raton, FL.

Grant Maxie, M. and Robinson, W.F. (2007) Cardiovascular system. In: Grant Maxie, M. (ed.). Jubb, Kennedy and Palmer's Pathology of Domestic Animals, vol. 2, Saunders, Philadelphia, p. 71.

Hardisty, J.F. (1985) Factors influencing laboratory animal spontaneous tumor profiles. *Toxicologic Pathology*, 13, 95–104.

Hartman, L.A. (1987) Idiopathic extramural coronary arteritis in beagle and mongrel dogs. *Veterinary Pathology*, 24, 537–44.

Henderson, K.S., Dole, V., Parker, N.J., Momtsios, P., Banu, L., Brouillette, R., Simon, M.A., Albers, T.M., Pritchett-Corning, K.R., Clifford, C.B. and Shek, W.R. (2012) Pneumocystis carinii causes a distinctive interstitial pneumonia in immunocompetent laboratory rats that had been attributed to 'rat respiratory virus'. *Veterinary Pathology*, 49, 440–52.

Jeppesen, G. and Skydsgaard, M. (2015) Spontaneous background pathology in Göttingen minipigs. *Toxicological Pathology*, 43(2), 257–66.

Johnson, R.C., Spaet, R.H. and Potenta, D.L. (2013) Spontaneous lesions in control animals used in toxicity studies. In: Sahota, P.S., Popp, J.A., Hardisty, J.F. and Gopinath, C. (eds). Toxicologic Pathology: Nonclinical Safety Assessment, CRC Press, Boca Raton, FL, pp. 209–57.

Keenan, K.P., Coleman, J.B., McCoy, C.L., Hoe, C.M., Soper, K.A. and Laroque, P. (2000) Chronic nephropathy in ad libitum overfed Sprague-Dawley rats and its early attenuation by increasing degrees of dietary (caloric) restriction to control growth. *Toxicologic Pathology*, 28(6), 788–98.

Lewis, D.J. and McKevitt, T.P. (2013) Respiratory system. In: Sahota, P.S., Popp, J.A., Hardisty, J.F. and Gopinath, C. (eds). Toxicologic Pathology: Nonclinical Safety Assessment, CRC Press, Boca Raton, FL, pp. 367–421.

Livingston, R.S., Besch-Williford, C.L., Myles, M.H., Franklin, C.L., Crim, M.J. and Riley, L.K. (2011) Pneumocystis carinii infection causes lung lesions historically attributed to rat respiratory virus. *Comparative Medicine*, 61, 45–52.

Long, G.G. and Hardisty, J.F. (2012) Regulatory forum opinion piece: thresholds in toxicologic pathology. *Toxicologic Pathology*, 40, 1079–81.

MacKenzie, W.F. and Alison, R.H. (1990) Heart. In: Boorman, G.A., Eustis, S.L., Elwell, M.R. and MacKenzie, W.F. (eds). Pathology of the Fischer Rat, Academic Press, San Diego, CA, pp. 461–71.

McInnes, E.F. (2012a) Preface. In: McInnes, E.F. (ed.). Background Lesions in laboratory Animals, Saunders Elsevier, Edinburgh, pp. vii.

McInnes, E.F. (2012b) Artifacts in histopathology. In: McInnes, E.F. (ed.). Background Lesions in Laboratory Animals, Saunders Elsevier, Edinburgh, pp. 93–101.

McInnes E.F. (2012c) Minipig. In: McInnes, E.F. (ed.). Background Lesions in Laboratory Animals, Saunders Elsevier, Edinburgh, pp. 81–7.

McInnes, E.F. and Scudamore, C.L. (2014) Review of approaches to the recording of background lesions in toxicologic pathology studies in rats. *Toxicology Letters*, 229, 134–43.

McInnes, E.F. and Scudamore, C.L. (2015) Aging lesions: background versus phenotype. Current Pathobiology Reports, doi:10.1007/s40139-015-0078-y.

Montgomery, C.A. and Seely, J.C. (1990) Kidney. In: Boorman, G.A., Eustis, S.L., Elwell, M.R. and MacKenzie, W.F. (eds). Pathology of the Fischer Rat, Academic Press, San Diego, CA, pp. 127–52.

Mori, H. and Matsumoto, K. (1973) Development of the secondary interstitial gland in the rabbit ovary. *Journal of Anatomy*, 116, 417–30.

Percy, D.H. and Barthold, S.W. (2007) Pathology of Laboratory Rodents and Rabbits, 3rd edn, Blackwell, Ames, IA, pp. 3–124.

Rehm, S. (2000) Spontaneous testicular lesions in purpose bred beagle dogs. *Toxicologic Pathology*, 28, 782–7.

Son, W.C. (2003) Factors contributory to early death of young CD-1 mice in carcinogenicity studies. *Toxicology Letters*, 145, 88–98.

Son, W.-C. (2004a) Idiopathic canine polyarteritis in control beagle dogs from toxicity studies. *Journal of Veterinary Science*, 5, 147–50.

Son, W.C. (2004b) Factors contributory to death of young Sprague-Dawley rats in carcinogenicity studies. *Toxicology Letters*, 153, 213–19.

Son, W.C. and Gopinath, C. (2004) Early occurrence of spontaneous tumors in CD-1 mice and Sprague-Dawley rats. *Toxicologic Pathology*, 32(4), 371–4.

Son, W.C., Bell, D., Taylor, I. and Mowat, V. (2010) Profile of early occurring spontaneous tumors in Han Wistar rats. *Toxicologic Pathology*, 38, 292–6.

Sundberg, J.P., Taylor, D., Lorch, G., Miller, J., Silva, K.A., Sundberg, B.A., Roopenian, D., Sperling, L., Ong, D., King, L.E. and Everts, H. (2011) Primary follicular dystrophy with scarring dermatitis in C57BL/6 mouse substrains resembles central centrifugal cicatricial alopecia in humans. *Veterinary Pathology*, 48, 513–24.

Suttie, A.W. (2006) Histopathology of the spleen. *Toxicologic Pathology*, 34, 466–503.

Swindle, M.M., Makin, A., Herron, A.J., Clubb, F.J. Jr and Frazier, K.S. (2012) Swine as models in biomedical research and toxicology testing. *Veterinary Pathology*, 49, 344–56.

Weber, K., Mowat, V., Hartmann, E., Razinger, T., Chevalier, H.J., Blumbach, K., Green, O.P., Kaiser, S., Corney, S., Jackson, A. and Casadesus, A. (2011) Pathology in continuous infusion studies in rodents and non-rodents and ITO (infusion technology organisation) – recommended protocol for tissue sampling and terminology for procedure-related lesions. *Journal of Toxicologic Pathology*, 24(2), 113–24.

5

Target Organ Pathology

Elizabeth McInnes

Cerberus Sciences, Thebarton, SA, Australia

Learning Objectives

- Recognise the major macroscopic pathology findings in the skin, gastrointestinal system, liver and other tissues.
- Understand the macroscopic pathology lesions caused by the major xenobiotic compounds.
- Understand renal and hepatic failure.
- Understand phospholipidosis.

   In theory, all tissues and organs in the body are potential targets for the toxic effects of xenobiotics. In practice, however, certain organs are affected more than others. The skin, eyes, respiratory tract and gastrointestinal tract are often target organs as a result of exposure due to the route of administration (intradermal, inhalation, oral gavage, etc.). In addition, tissues associated with metabolism and excretion, such as the liver and kidney, often demonstrate test article-related findings. Generally, the responses to xenobiotic toxicity in all tissues follow the same form as those induced by other causes: degeneration, necrosis, acute and chronic inflammation, proliferation and neoplasia, as discussed in Chapter 3. The assessment of the toxicity of a xenobiotic is determined by establishing whether it elicits an effect in more than one animal species, whether there is a clear dose response between the low- and high-dose animal groups and, in the case of tumours, whether they are benign or malignant, or both. In this chapter, emphasis is placed on lesions that may be observed by study personnel at necropsy, rather than complex histopathological lesions only visible under the light microscope.

5.1   Skin

The skin is the external covering of the body. As one of the largest organs of the body, it is vulnerable and often injured. Changes to the skin and fur may be the most obvious changes seen in laboratory animals on study. The presence of abundant fur in most species (with the exception of the minipig), and the associated difference in skin

Pathology for Toxicologists: Principles and Practices of Laboratory Animal Pathology for Study Personnel,
First Edition. Edited by Elizabeth McInnes.

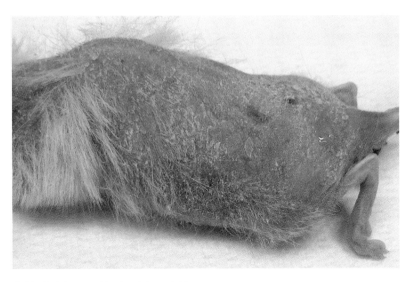

Figure 5.1 Fur/hair loss or thinning (alopecia) in a mouse.

physiology, means that any changes seen must be carefully evaluated for their potential relevance to humans. A fur covering may also mask changes to the underlying skin, which can only be visualised by microscopic examination. Changes in the skin may be seen both following systemic administration of drugs (although this is quite rare) and as a result of topical drug application (more common). In addition, the skin is capable of biotransformation and may produce toxic metabolites from a harmless compound (Oesch et al., 2007).

Fur/hair loss or thinning (Figure 5.1) may occur as a result of systemic administration of drugs, and reflects changes in hair growth or follicular destruction. Loss of hair (alopecia), reduced hair production (hypotrichosis) and excessive hair production (hypertrichosis) may all be observed after topical or systemic administration of xenobiotics. Alopecia is caused by a large number of drugs, including cytotoxic drugs and retinoids (Haschek et al., 2010). Topical xenobiotics may cause systemic effects, or a xenobiotic administered systemically may have effects on the skin (e.g. corticosteroids administered orally may cause skin thinning (atrophy) in dogs).

Fur/hair colour may change with the administration of xenobiotics. Melanocytes (which are black) occur in the skin of pigmented animals. Oestrogens tend to cause an increase in skin pigmentation (i.e. a dark brown black colour) (hyperpigmentation), whilst androgens cause a decrease in pigmentation (hypopigmentation) (Tadokoro et al., 2003). Some nanoparticles can cross the skin and enter the lung and nerves, but nanocrystalline silver in wound dressings localises only in the superficial layers of the stratum corneum, causing grey discolouration (Samberg et al., 2010).

Skin reddening (erythema) and scaling may be most obvious where areas of skin have been exposed (e.g. via hair clipping) to allow topical administration of drugs. This represents the early inflammatory response to irritation (i.e. vasodilatation of blood vessels and skin thickening). It can be induced by the method used to remove the fur (e.g. shaving or chemical depilation), and it is important that this factor is controlled for by preparing sites in the same way and then applying vehicle at a control site. Vasodilation causing erythema and blister or vesicle formation (accumulation of oedematous fluid) may also be observed at necropsy, after exposure to irritant xenobiotics such as toluene. Fibrosis and scarring will result after initial erosion and ulceration of the skin.

Skin thickening (epidermal thickening is called 'acanthosis') (Figure 5.2) and scaling may be hard to appreciate if the fur is intact, but may be noted in life as an increase in loose scale ('dandruff'). It is often associated with modification of sebaceous gland activity. Skin erosion (partial loss of the epidermis) and ulceration (complete loss of the epidermis) may be easily recognised as red craters lined by yellowish crusts visible in the overlying skin (Figure 5.3). Ulceration occurs as a result of direct necrosis of the skin tissue caused by corrosive agents or as a secondary effect of vasculitis, leading to local ischaemia of the skin. Compounds that cause ulceration include irritant substances such as acids and alkalis, as well as drugs that cause immune-mediated responses, vasculitis and/or photosensitivity reactions (including sulphonamides, penicillins and anticonvulsant drugs) (Haschek et al., 2010). Photosensitivity is noted in the skin when ultraviolet (UV) light activates photosensitive compounds (e.g. fluoroquinolones, tetracyclines and retinoids) (Ferguson, 2002) present either on its surface (topical administration) or within it (systemic administration), causing erythrema, blistering and eventual sloughing of the skin.

Pustule and abscess formation generally occurs as a result of inflammation of the hair follicle. Pustules are easy to recognise, generally consisting of a 'pimple-like' structure with a central mass of pus (yellow) surrounded by a layer of fibrosis or scar tissue. Inflammation of the hair follicle may cause the formation of pustules, and later micro- and macroabscesses. Acne can be caused by chlorinated hydrocarbons and is characterised by dilated sebaceous glands filled with keratin.

Figure 5.2 Acanthosis (epidermal thickening) in a young rat pup.

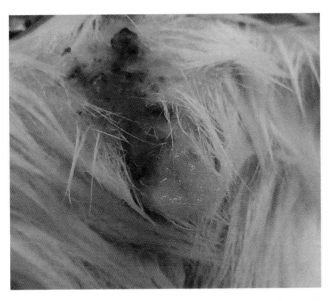

Figure 5.3 Necrosis, erosion and ulceration of the skin in the preputial gland area of a male mouse.

Figure 5.4 Petechiae on the skin of a nude mice.

Petechiae are small (~1–2 mm diameter) red haemorrhages in the skin or subcutis (Figure 5.4). They are easy to see in nude mice. They may be caused by systemic compounds (e.g. sulphonamides), eliciting widespread vasculitis (inflammation of blood vessels) or disseminated intravascular coagulation (widespread microthrombi lodging in vessels, causing macroscopic petechiae) (Wojcinski et al., 2013).

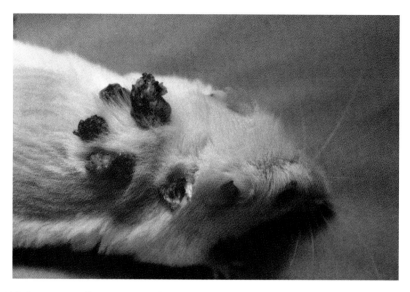

Figure 5.5 Squamous cell carcinoma in the dorsal skin above the scapula in a rat.

Granulomas are a distinct reaction in various tissues (e.g. lungs and skin), causing focal, raised masses. They typically contain a centre made up of foreign material or microorganisms, surrounded by large multinucleate macrophages, neutrophils and a layer of fibrosis. Lesions caused by drugs in the dermis include granulomatous inflammation due to modified hyaluronan (Westwood et al., 1995) or a foreign body, such as a medical device. Calcification or mineralisation of the subcutis may be observed on treatment with substances such as dihydrotachysterol (Wojcinski et al., 2013), causing the skin to become hard and gritty when cut.

Skin tumours may involve the epidermis (e.g. squamous cell papilloma, squamous cell carcinoma; Figure 5.5) or may be derived from mesenchymal cells (e.g. fibrosarcoma, lipoma, mast cell tumour, haemangiosarcoma). Tumours of the hair follicles include trichofolliculoma and trichoepithelioma. Melanocytic tumours (generally brown or black in colour) of the skin include benign and malignant melanoma; they may be induced by UV light (Ingram, 1998). Chronic administration of an irritant xenobiotic to the skin can cause the development of tumours, such as squamous cell carcinoma (generally, a large, ulcerated, raised skin mass that does not heal). Biomaterials such as microchips may also be carcinogenic, and may cause mesenchymal tumours in the skin due to chronic irritation (Elcock et al., 2001).

5.2   Eye

Injury to the eye occurs in all species of laboratory animals, and ranges from minimal (e.g. conjunctivitis (Figure 5.6), where only the skin round the eye is affected) to severe (i.e. panophthalmitis, where all are components are inflamed). Bulging eyes are easy to recognise in all laboratory animal species; this phenomenon is termed 'exophthalmos' and is generally secondary to inflammation of the orbit or the presence of tumours

Figure 5.6 Conjunctivitis in a mouse.

(e.g. a Harderian gland adenoma or adenoacarcinoma behind the eye). Excessive tear production (lacrimation) and brown discolouration (red tears) of the periorbital skin may be observed, particularly in rodents. Excessive lacrimation can be seen around the eyes and is caused by stress or cholinergic drugs (Harkness and Ridgeway, 1980).

Dry eyes with a reduction in tear production leads to redness and scaling in the conjunctiva; this is known as 'keratoconjunctivits sicca'. Radiation that causes degeneration of the lacrimal gland can lead to a reduction in tears and as well as local anaesthetics. Cloudiness of the cornea tends to indicate keratitis (inflammation of the cornea) and is caused by dust, irritant substances and topical administration of prostaglandin agonists (Aguirre et al., 2009). Initially, keratitis presents as a cloudiness to the cornea (oedema); later, ulceration can occur. An injured eye may heal with fibrosis (scarring), causing a white cloudiness of the cornea (pannus).

Inflammation of the skin around the eye is called 'conjunctivitis', and the periorbital skin will be red and swollen (oedema), and may be ulcerated and weeping. Causes include infectious agents and compounds such as ricin (Strocchi et al., 2005). The accumulation of blood in the anterior chamber is called 'hyphema', whilst the accumulation of pus in the anterior chamber is called 'hypopyon'; this is seen in cases of bacterial infection (e.g. *Staphylococcus aureus*). Inflammation of the iris (iriditis) may resolve with fibrosis and can be caused by cyclophosphamide. Impaired drainage of the aqueous humour is referred to as 'glaucoma'.

The lens of the eye is normally transparent. 'Cataract' is the degeneration of the lens, causing whitish opacification of the eye (Figure 5.7) and a white-appearing lens. Cataract may be caused by age, diabetes, UV light or compounds such as glucocorticoids. Retinal atrophy is common in older mice and may be caused by ageing or continuous lighting. It is not possible to see retinal atrophy without performing an ophthalmoscopic examination (fundoscopy) or looking at the retina under the light microscope.

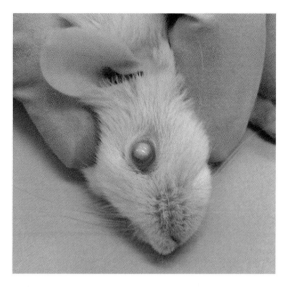

Figure 5.7 Whitish opacification of the eye due to the presence of a cataract.

5.3   Gastrointestinal Tract

The gastrointestinal tract is one of the largest organs of the body. It is a prime target for test article-related lesions because it is the first organ in contact with the test substance following ingestion. Gastric lavage and oral ingestion are two of the most common routes of chemical administration, with a high likelihood of intestine-related toxicity. Clinical signs relating to the gastrointestinal tract are obvious and easily recognised in laboratory animals on study; these include diarrhoea, salivation, blood in the faeces and vomiting in some animals (but not rats or mice). The critical balance between constant cell division and cell loss in the stomach and small and large intestines means that ulceration and proliferation occur commonly. In general, the mechanism of injury in the gastrointestinal tract is irritation, and the response is degeneration/necrosis, inflammation and proliferation.

Thickened gums and tooth abnormalities may be observed in the mouths of laboratory animals on study. Calcium channel blockers such as cyclosporine A cause gingival hyperplasia (excessive thickening of gums) in dogs, whilst immunosuppressants such as corticosteroids may cause ulceration with subsequent *Candida* overgrowth due to the inability of the immune system to defend against the invasion of yeasts. Tetracyclines cause tooth discolouration, and dental abnormalities are noted with cyclophosphamide treatment. Vascular endothelial growth factor (VEGF) is an angiogenesis inhibitor and may cause dental dysplasia (disordered growth) and degeneration resulting in tooth fractures, which may be visible at necropsy as broken and extremely white teeth. Sunitinib, a broad-spectrum tyrosine kinase inhibitor, causes gingival necrosis in non-human primates (Patyna et al., 2008), which is visible as focal areas of ulceration on the gums. Tumours of the mouth include squamous papilloma (Figure 5.8) and squamous cell carcinoma of the oral mucosa. Tumours of the teeth include odontoma, fibroma and ameloblastoma; these may occasionally be induced by compounds such as 3-methylcholanthrene (Greene et al., 1960).

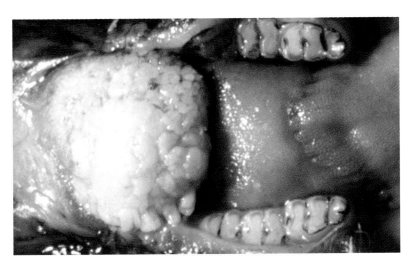

Figure 5.8 Squamous papilloma in the mouth of a rat.

Excessive salivation is characterised by increased amounts of saliva. It is easy to recognise in dogs and non-human primates, with strands of saliva extending from the mouth over the chin. Beta adrenergic agonists may cause salivary-gland enlargement (Ten Hagen et al., 2002), and excessive salivation is seen in organophosphate toxicity (Betton, 1998a). Salivary gland inflammation and atrophy due to reduced food consumption is a common test article-related finding. Tumour development in the salivary glands includes adenoma, adenocarcinoma, myoepithelioma and mesenchymal tumours and is rarely caused by administration of xenobiotics.

Difficulty in swallowing can be observed in rodents, dogs and non-human primates, and may be related to changes in the oesophagus. In these cases, undigested food will be present either in the mouth or on the floor in front of the affected animal. Lesions of the oesophagus include oesophagitis and gavage injury (where the gavage tube goes through the oesophageal tissue into the thoracic cavity and causes inflammation). Oesophagitis generally heals with scar formation and fibrosis, and obstruction may result (Betton, 1998a). Megaoesophagus (enlarged dilated oesophagus) causes difficulty in swallowing and is generally congenital. Zinc and vitamin A deficiencies can cause thickening of the oesophagus (hyperkeratosis), which is visible as a thickened white internal oesophageal lining. Oesophageal tumours include papilloma and carcinoma and may be induced by nitrosaniline and nitrosourea compounds in rats (Markovits et al., 2013).

Stomach ulceration and vomiting are often observed when the lining of the stomach is damaged. During life, the animal may show a hunched position, indicating abdominal pain. Stomach ulcerations can be easily recognised at necropsy, and consist of small reddened craters (full- or partial-thickness ulceration of the mucosa) present in the glandular and nonglandular stomach of rodents or in the glandular stomach of dogs. Ulcers in the stomach may lead to gastric haemorrhage, which is characterised by dark black gastric contents as the blood is digested by the gastric enzymes. Irritant compounds, such as ethanol, acids and alkalis (Betton, 1998a), cause initial erosion and ulceration, and later can also cause proliferative changes and thickening of the mucosa of the stomach. Mineralisation of the stomach is unlikely to be seen at necropsy, although it may manifest as a grittiness during cutting of the stomach with a knife or scalpel at necropsy. Mineralisation of the mucosa and muscle layers of the stomach may be observed with MEK inhibitors (Diaz et al., 2012), but this is a rare lesion.

Thickening of the stomach wall can be recognised at necropsy, especially when treated stomachs are compared to the stomachs of control animals. Hyperplasia of the glandular stomach (particularly the neuroendocrine cells of the stomach) progressing to neoplasia may be observed with antisecretory drug treatment, such as omeprazole (neuroendo-crine cell proliferation), as well as with irradiation. Antisecretory drugs such as cimetidine restrict the production of hydrochloric acid in the stomach, stimulating the neuroendocrine cells to increase and enlarge. Stomach tumours are easy to recognise, generally consisting of large, raised, red, ulcerated masses. They may be induced by xenobiotics such as nitosamines. Tumours in the stomach include papilloma, squamous cell carcinoma, adenoma, adenocarcinoma, neuroendocrine tumours and mesenchymal tumours. Squamous cell carcinoma is commonly induced by irritants, promoters and genotoxins (Figure 5.9) (Chandra et al., 2010).

Diarrhoea is common in laboratory animals, and pools of liquid (often mucoid or bloody faeces) are generally visible in the caging. At necropsy, the small and large intestines will be moderately dilated and filled with foul-smelling green, brown or red liquid contents. In severe cases, diarrhoea can result in the telescoping of the small intestine into a downstream section, causing a lesion known as 'intussusception'; this may be caused by compounds such as alpha adrenergic agonists, which increase intestinal motility and thus cause diarrhoea. The enterocytes of the intestinal surface have a high mitotic rate and are thus very susceptible to cytotoxic drugs such as cyclosporine, as well as radiation. Irritants compounds (such as heavy metals) cause mucosal ulceration, haemorrhage and inflammation of the intestine, with consequent diarrhoea. Thickening of the intestine may be related to proliferative lesions in the intestine, caused by drugs such as VEGF receptor inhibitors, which lead to proliferation of Brunner's glands in the duodenum (Ettlin et al., 2010).

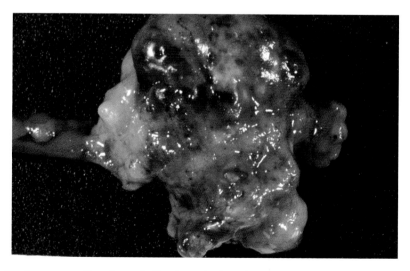

Figure 5.9 Squamous cell carcinoma in the stomach of a rodent.

The large intestine may show evidence of diarrhoea or constipation. Occasionally, the intestine may display a different colour at necropsy. Some macromolecules derived from drug compounds may accumulate in macrophages within the lamina propria of the small or large intestine or the regional lymph nodes, causing discolouration. Xenobiotics can interfere with motility in the large intestine, causing either diarrhoea or constipation (morphine derivatives), and antibiotics such as tetracyclines change the intestinal microflora, resulting in diarrhoea. Caecal enlargement is associated with increased calcium absorption or ingestion of dietary starches and sugar alcohols, as well as nonabsorbable or high-molecular-weight compounds (Betton, 1998a). Thickening of the small or large intestine may indicate underlying inflammation (i.e. enteritis). Chronic colitis in the mouse is characterised by thickening of the colon and rectum, which is noticeable at necropsy (Figure 5.10). Intestinal tumours are rare, but include polyps (Figure 5.11), adenoma, adenocarcinoma and mesenchymal tumours, such as leiomyoma and leiomyosarcoma (Figure 5.12). Peritonitis is inflammation of the lining of the abdominal cavity and is characterised by yellow strands of fibrin and pus and by turbid fluid in the abdominal cavity at necropsy. Peritonitis is rare in toxicological studies, but may occur if a treatment-related ulcer in the stomach or the small or large intestine perforates and faeces enters the abdominal cavity.

The pancreas has two elements: the exocrine and the endocrine pancreas (see Section 5.10). Pancreatitis (inflammation of the exocrine pancreas) is an acute and severe condition, and animals will display severe pain (e.g. hunched posture in rodents, the prayer position in dogs). At necropsy, the pancreas will display severe haemorrhage, necrosis (yellow areas) and oedema. Pancreatitis may be caused by drugs such as cyclosporine A (Hirano et al., 1992), but not many compounds are known to cause it. Increased or decreased zymogen granules in the pancreas (only visible under the light

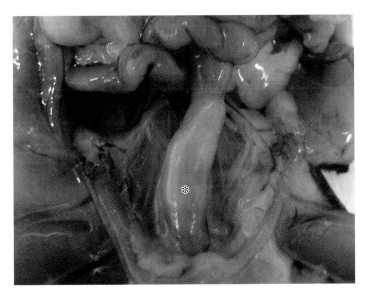

Figure 5.10 Chronic colitis (*) in a mouse.

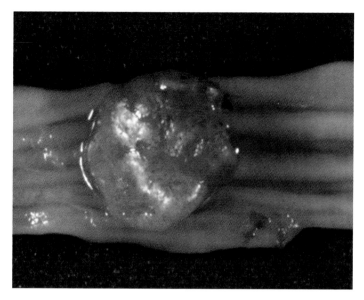

Figure 5.11 Polyp in the colon of a rat.

Figure 5.12 Leiomyosarcoma extending from the ileum outer surface in a rat.

microscope) may be caused by reduced feed consumption or sunitinib, an inhibitor of receptor tyrosine kinase (Patyna et al., 2008); this is a frequent finding in short- and long-term safety studies.

Proliferative lesions are visible as small to large masses in the pancreas at necropsy. Proliferative lesions in the pancreas include acinar hyperplasia, adenomas and carcinomas; these may be observed with peroxisome proliferator-activated receptor (PPAR) alpha activators, fibrates (Cattley et al., 2013), as well as heat-labile, soya-bean, trypsin inhibitors (Betton, 1998a).

5.4   Liver

The liver is a common target organ in toxicity studies in animals because oral compounds reach it first via the hepatic portal vein from the gastrointestinal tract. Furthermore, the liver is often responsible for removing all of a compound from the blood, and is thus exposed to the compound in high concentrations. Toxic liver injury is one of the most frequent reasons for terminating the development or therapeutic use of a drug. Test article-related findings in the liver include degeneration, necrosis and regeneration. Liver failure is characterised by jaundice, encephalopathy (neurologic signs such as dullness), bleeding tendency (due to deficiency of clotting factors), odema and hypoproteinaemia.

Liver necrosis and inflammation are common, and are recognised at necropsy by the presence of small to large, nonraised yellow areas present on the reddish-brown surface of the liver (Figure 5.13). Hepatocyte degeneration or necrosis may range from a single cell (only observed under microscope) to multifocal, lobar and massive necrosis. Test article-related liver necrosis is a significant lesion in toxicology, and its presence may mean that a compound cannot proceed to clinical trials (Cattley et al., 2013). Common mechanisms of liver necrosis include depletion of glutathione, interference with mitochondrial energy production and cytoskeleton damage (Grattagliano et al., 2009). Liver necrosis is generally either centrilobular (the most common form, due to the concentration of CYP enzymes in this area and the anoxia that occurs here, at the furthest point from the arteriole), midzonal or periportal. These distributions can only be observed under light microscope, but will often be mentioned in pathology reports as they describe the injury in terms of major liver landmarks (i.e. the portal tract and central vein). If the necrotic injury involves the mesenchymal tissue in the liver, scarring or fibrosis will occur; cirrhosis is defined as a 'fibrosing nodular liver'.

Hypertrophy or enlargement of the hepatocytes (increase in size, not in number) is a common toxicological liver lesion. It may be observed at necropsy by the presence of a large liver (with increased weight compared to controls) with rounded edges that

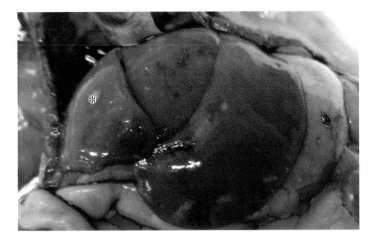

Figure 5.13 Yellow areas (*) on the reddish-brown surface of the liver, indicating liver necrosis and inflammation.

bulge upon cutting. Hypertrophy of hepatocytes is generally associated with an increased cytoplasmic endoplasmic reticulum due to P450 enzyme induction or peroxisome proliferation. Noted peroxisome proliferators are the hypolipidaemic agents based on clofibric acid and the PPARs. Xenobiotic enzyme inducers (e.g. phenobarbitone) produce enlarged hepatocytes capable of rapidly degrading endogenous hormones such as sex hormones and thyroid-stimulating hormone (TSH). This may result in thyroid hypertrophy, as the thyroid responds to the greater TSH breakdown by producing more TSH. It is thus a good idea to examine the thyroids carefully in animals that display an enlarged liver, as they may be increased in size at necropsy.

A diffuse pale yellow is often seen at necropsy. Lipid accumulation is one of the most common treatment-related findings in the liver, and the liver will generally be increased in weight and may assume a yellow colour and greasy texture. Oil Red O and Sudan black are special stains that can be used on histological sections of the liver (nonfixed) to identify lipid in the liver tissue. Accumulation of pigments (lipofuscin, glycogen) or nonpigments (drug metabolites) may also be a test article-related finding, and may cause a colour change in the liver, which will be visible at necropsy. For instance, hypolipidaemic agents cause an increase in lipofuscin (golden-brown pigment) in rodent liver cells following prolonged treatment (increased lipid peroxidation).

Hepatitis is inflammation of the liver. At necropsy, a swollen, enlarged, often pale yellow or mottled liver can be seen, frequently with thin yellow fibrin strands on the surface. This change is not common as a test article-related finding. Inflammatory cell infiltration into the liver is a common background lesion and is generally only visible under light microscope, but it may rarely be a test article-related finding (e.g. atorvastatin in beagle dogs in chronic toxicity studies causes hepatic microgranulomas). The inflammatory-cell aggregations may be visible macroscopically as tan, white foci in the liver, if severe. Immunosuppressive drugs can cause a decrease in inflammatory cell foci in the liver. Certain xenobiotics may be phagocytosed by Kupffer cells (macrophages of the liver), including nanoparticles (Xiao et al., 2011); occasionally, this may be visible as a colour change in the liver at necropsy.

A generalised yellow colour of the fat tissues and carcase at necropsy is referred to as 'jaundice' or 'icterus' (Figure 5.14); this may have prehepatic (generally, red blood cell breakdown), heaptic (such as liver inflammation) or posthepatic (obstruction of the large bile duct, preventing the bile from moving to the gall bladder) causes. Examination of the sclera of the eye is a good method for establishing that jaundice is present. Cholestasis is a decreased volume of bile or impaired secretion of specific solutes into the bile, with a resultant increase in the serum levels of bile salts and bilirubin. Icterus may also be caused by the drug chlorpromazine or phalloidin (Betton, 1998b). Test article-related findings in the gall bladder and bile duct include inflammation (cholecystitis), bile-duct necrosis (caused by the compound trabectidin) (Donald et al., 2002) and bile-duct proliferation.

Liver tumours are generally visible at necropsy as multiple or single masses, often raised above the liver surface, and sometimes round. Hepatic proliferative lesions caused by xenobiotics range from hypertrophy and hyperplasia to hepatocellular adenomas (Figure 5.15) and hepatocellular carcinomas. Foci of cellular alteration (esoinophilic, basophilic, vacuolated and mixed) are not visible macroscopically, but may be test article-related; controversy remains over whether or not these lesions are preneoplastic (Thoolen et al., 2013). Tumours of the bile duct include cholangioma (benign) and cholangiocarcinoma and hepatocholangiocarcinoma (malignant); these generally

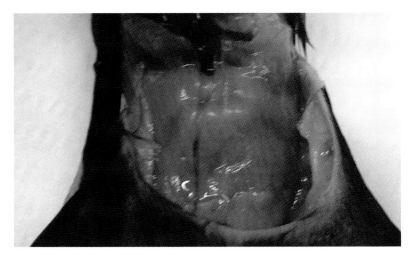

Figure 5.14 Jaundice/icterus in a mouse, visible as a generalised yellow colour of the abdominal muscles.

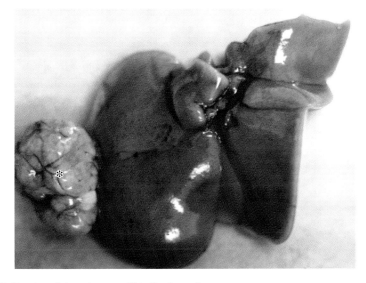

Figure 5.15 Hepatocellular adenoma (*) in the liver of a mouse.

cause jaundice, because the bile is prevented from leaving the gall bladder by the obstruction of the tumour in the bile duct.

5.5 Respiratory System

It is difficult to visualise the respiratory system whilst animals are living, although the respiration rate will give you lots of information about what is happening in the lungs (an increased respiration rate is seen in cases of pneumonia or lung tumours). A nasal discharge (clear, mucoid or pus) gives an indication of inflammation in the upper respiratory tract. The respiratory system is analysed in detail in inhalation studies, although compounds administered by other routes (such as oral gavage) may also effect the respiratory epithelia via the bloodstream or through inhalation of refluxed material (Lewis and McKevitt, 2013). Generally, in inhalation studies, four sections of the nasal cavity are examined, and three or four standard sections of the rodent nasal turbinates may be cut with reference to the upper palate (Young, 1981). Local metabolism of inhaled or administered compounds occurs in the nasal epithelia and Bowman's glands, due to the presence of xenobiotic metabolising enzymes such as P450 (Harkema, 1991). Mice and rats are obligate nose-breathers, and so obstruction to airflow due to inflammation or a tumour may result in swallowing of air, leading to a severely distended, air-filled stomach and intestines, which eventually compresses the diaphragm, causing death by asphyxiation (Figure 5.16).

Inflammation in the nasal tract may be serous, fibrinous or mucous, generally resulting in a watery, mucoid or purulent nasal discharge. Irritation is a common cause of upper respiratory-tract lesions. Here, the respiratory tract will be red and oedematous, although this may not be visible at necropsy, since the upper respiratory tract is often not opened. Thus, test article-related findings in the upper respiratory tract are generally not observed at necropsy, since the actual epithelia lining the nasal turbinates are not examined and the head is fixed and decalcified whole. Regeneration of damaged epithelium may also include metaplasia, where nasal epithelium transforms into a more resistant type, such as squamous epithelium. Severe epithelial necrosis and inflammation of the upper respiratory tract may cause turbinate fusion, as repair and fibrosis occur after the initial insult, and this will manifest during life as difficulty in breathing. Test article-related findings in the nasal epithelia of the upper respiratory tract (i.e. nasal turbinates, larynx, trachea and bronchi) include atrophy, hypertrophy, hyperplasia,

Figure 5.16 Distended, air-filled murine stomach and intestines, due to obstruction of the airflow due to inflammation of the nasal turbinates.

degeneration, necrosis, erosion and ulceration. Test article-related tumours in the nasal cavity are rare and tend to be papillomas, adenomas or adenocarcinomas and carcinomas. Common test article-related findings in the larynx include degeneration and necrosis of the epithelium of the ventrolateral areas, and ventral pouch and squamous epithelium of the arytnoid cartilages. Cilia loss and epithelial degeneration are seen in the trachea, and particularly at the tracheal bifurcation, which is a common area of damage in inhalation studies.

Xenobiotic-associated changes may also be observed in the nasal-associated lymphoid tissue (NALT) and in the underlying blood vessels of the upper respiratory tract, and corticosteroids in particular are known to cause depletion of lymphocytes in dog NALT (Haley, 2003). Clara cells are common in the trachea epithelium, where they function to produce surfactant (Suarez et al., 2012) and may respond to inhaled compounds. Identification of Clara cells is generally performed using immunohistochemistry, where specific antibodies are used to identify them. Inhaled corticosteroids cause enlargement (hypertrophy) of mouse Clara cells (Roth et al., 2007). All of these findings may be visualised only under the light microscope.

Macrophage reactions in the lung remain one of the most common findings noted with inhaled pharmaceuticals, since macrophages are very effective at phagocytosing inhaled material. At necropsy, slightly raised white foci are visible on the lung surface, indicate the presence of macrophage aggregates, such as those seen in PPAR alpha agonists (Figure 5.17). The presence of alveolar macrophages in the lung alveoli is a background change in rats and mice, so interpretation of drug-induced alveolar macrophage increase may be difficult. A significant cause of attrition during the development of new inhaled drugs is the induction of foamy macrophage aggregates within the lungs, particularly in rats (Lewis et al., 2014). This generally occurs because the inhaled compounds are made poorly soluble in order to reduce systemic absorption.

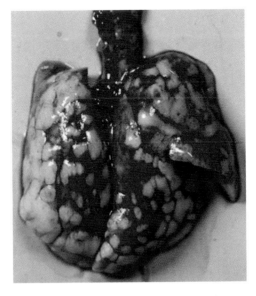

Figure 5.17 Slightly raised white foci on the lung surface, indicating the presence of macrophage aggregates.

Spontaneous alveolar macrophage aggregates are generally observed randomly dispersed throughout the lung tissue, whereas the induced macrophage responses tend to be located at the bronchioalveolar junctions (Lewis and McKevitt, 2013). Alveolar macrophages may ingest pigments (e.g. carbon), and this can cause the presence of blackish areas on the pleural surface at necropsy. Latterly, researchers consider that macrophage aggregates (either spontaneous or induced) present only at bronchioalveolar junctions are harmless, as they are reversible (Lewis et al., 2014). These macrophage aggregates are not associated with cell damage, do not progress and are not associated with any other type of inflammatory cell (Lewis et al., 2014). Type 2 macrophage aggregates with evidence of infiltrating neutrophils or lymphocytes and type 3 reactions with granulomatous inflammation are irreversible and are considered to be adverse.

Test article-related findings in the lungs include alveolar type II cell hypertrophy and hyperplasia, as well as acute and chronic inflammatory reactions induced by immunogenic proteins (Clarke et al., 2008). These changes are not visible at necropsy.

At necropsy, pneumonia will present with lungs that are hard and consolidated. The lungs in an acute bronchopneumonia tend to be red, whilst a chronic pneumonia is often yellow-white. Although pneumonia is a rare test article-related finding in the lungs of laboratory animals, it may be observed in animals treated with immunosuppressive compounds, such as corticosteroids. Pneumonia may be noted when aspiration of the stomach contents occurs. A foreign-body pneumonia may be seen with muscarinic antagonist treatment, since this entails an increased risk of oesophageal reflux; the lungs will often be brown-black in colour, with a fetid smell (Wilson and Walshaw, 2004). Silica dust causes progressive pulmonary fibrosis, which can be recognised at necropsy by the presence of hard nodules of scar tissue within the lung (Glaister, 1986). The most common xenobiotic-induced tumours of the lungs are bronchioloalveolar adenomas (Figure 5.18) and carcinomas, caused by compounds such as metronidazole (Contrera et al., 1997).

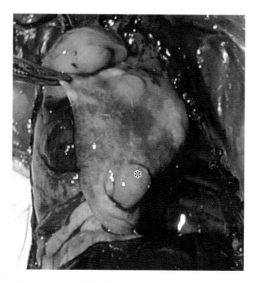

Figure 5.18 Bronchiolaveolar adenoma (*) in the lung of a mouse.

Phosholipidosisis is a common cause of foamy macrophage accumulation. It is produced by many drugs with cationic amphiphilic structures, which bind to hydrophobic or hydrophilic moieties of phospholipids, resulting in complexes that resist digestion by lysosomal phospholipase (Reasor et al., 2006). Using electron microscopy, single, membrane-bound, lysosomal bodies containing electron-dense smooth membranes arranged in stacks or as concentric whorls can be found in the cytoplasm. Using immunohistochemistry, positive staining for lysosome-associated protein 2 (LAMP-2) is noted in cases of phospholipidosis that are negative for Oil Red O and adipophilin. Many organs can be involved, displaying aggregates of large foamy macrophages. The liver may be involved, although lung tissue is a more common site. Phospholipidosis is considered to represent an adaptive phenomenon, rather than evidence of overt cell damage, but this remains controversial, with some toxicologists regarding it as adverse.

5.6   Urinary System

Kidney failure in most mammals is characterised by ulceration of the tongue, widespread oedema, metastatic calcification of tissues (e.g. stomach), hypertrophy of the parathyroids, nonregenerative anaemia (because the kidney is responsible for producing erythropoietin) and disorders of bone mineralisation (because the kidney plays a role in vitamin D synthesis). Dogs will have breath that smells of ammonia, and in general animals will lose weight due to loss of protein in the urine.

Tubular damage may not be noted at necropsy, but the presence of enlarged and light-brown or tan kidneys will be seen. Sometimes, however, tubular changes in the kidney are only visible under the light microscope. Compound-induced lesions are common in the kidney due to the high renal blood flow (25%) and high renal excretion of many drugs (Frazier and Seely, 2013). The renal tubular epithelial cells possess metabolic activity and may thus generate toxic metabolites, resulting in tubular damage. Tubular degeneration and consequent tubular regeneration and cortical tubular basophilia are common compound-induced changes often noted in the kidney when examined under the light microscope. Multifocal white to yellow nonraised spots or streaks on the kidney surface are suggestive of interstitial nephritis and may only be visible at necropsy when they involve large accumulations of lymphocytes into the cortex. Interstitial nephritis is noted commonly in the kidneys of mice, rats, dogs and non-human primates. This may be a background lesion in some species of laboratory animal (e.g. beagle dog), but it may also be test article-related generally if accompanied by further renal pathology and if a dose relationship can be demonstrated (Linton et al., 1980). Older rats are subject to chronic renal disease, which is exacerbated in male rats on a high-protein diet. The development of chronic progressive nephropathy in rats may be accelerated or retarded by various xenobiotics; for example, bromocriptine can slow the onset of renal disease in rats, whilst cyclosporine may exacerbate it (Greaves, 1998). Chronic progressive nephropathy or chronic interstitial nephritis is characterised by kidneys that are often smaller, shrunken, wrinkled and pale tan in colour at necropsy (Figure 5.19). Fibrotic tissue is nonelastic and causes contraction with consequent capsular depression. Decreased kidney weights are seen at necropsy in cases of longstanding renal disease and scarring.

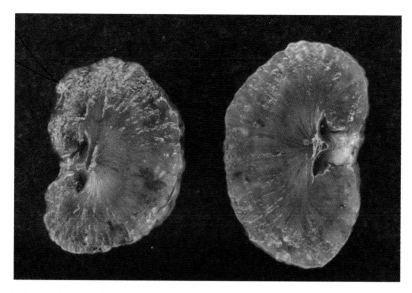

Figure 5.19 Shrunken, wrinkled, pale white rat kidneys at necropsy, indicating fibrosis and chronic interstitial nephritis.

Nephrotoxic agents often target specific parts of the renal nephron, and generally cause degeneration or necrosis of the tubular epithelial cells; these include gentamycin, heavy metals and cyclopsorin A. Compounds such as ethylene glycol produce oxalate crystals within the tubules, which cause tubular necrosis (Greaves, 1998), and melamine acts with a similar crystalline mechanism. Casts are only seen under the light microscope within the tubules; these include hyaline, protein casts, which indicate possible protein loss in the urine, and cellular casts, which contain inflammatory white blood cells (e.g. neutrophils) and indicate an infection of the kidney (pyelonephritis) (Figure 5.20). Pyelonephitis is characterised by pus within the pelvis and, occasionally, small yellow abscesses on the kidney surface. Bacterial cystitis is a predisposing factor for pyelonephritis; the lesion may be a test article-related finding caused by immuno-suppressive drugs such as cyclosporine (Remuzzi and Perico, 1995).

Glomerulonephropathy involves injury to the kidney glomerulus and is not generally visible at necropsy. Occasionally, in large animals such as dogs and minipigs, inflamed glomeruli may be visible in the kidneys at necropsy as small, red, pinpoint spots on the surface of the kidney. The glomerulus may be damaged directly or indirectly, and a common cause of glomerular damage is the aminonucleosoide of ouromycin, which causes damage to the glomerulus and increased permeability to plasma proteins, which eventually appear in the urine (proteinuria) (Glaister, 1986). Glomerular lesions may be induced by drugs, particularly biopharmaceuticals and macromolecules (Frazier and Seely, 2013), which often stimulate the production of antibodies and the activation of complement, resulting in immune complexes that lodge in the glomeruli and cause inflammation. These lesions are not visible at necropsy. Infarcts are clearly visible at necropsy, resulting in wedge-shaped, yellow areas sharply demarcated from the surrounding normal reddish renal tissue and involving both the cortex and the medulla of the kidney. Xenobiotics may cause renal infarcts if they initiate widespread thrombosis or if they are administered via an indwelling catheter or other forms of intravenous

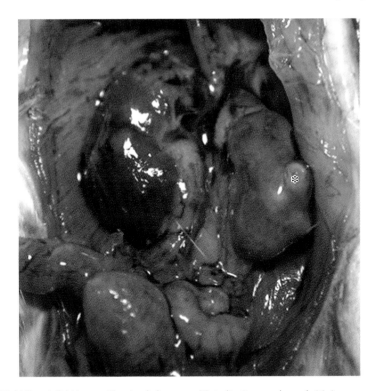

Figure 5.20 Yellow left kidney with raised abscesses (*), indicating pyelonephritis in a mouse.

infusion. Infarcts heal with the infiltration of scar tissue (fibrosis) into the ischaemic area. Amyloid is a pink material (positive with Congo red) that can only be diagnosed under the light microscope and which develops in dogs treated with auranofin (Greaves, 1998) and may be associated with chronic inflammation.

Hyaline droplets are eosinophilic, intracytoplasmic inclusions that generally occur in the cortical tubules and are only visible under the light microscope. They are thought to represent liposomes containing protein (Hard et al., 1999). Hyaline droplets normally occur in the mature male rat, where they represent reabsorption of alpha 2 microglobulin. Hyaline droplet nephropathy (i.e. accumulation of large numbers of hyaline droplets) can be induced by a large number of agents, including hydrocarbon and petroleum products and decalin. Xenobiotics may also induce the formation of hyaline droplets in the kidneys of female rats, and occasionally in mice.

Urinary stone formation (urolithiasis) is easy to recognise at necropsy and is characterised by the presence of crystals or small stones in either the bladder or the kidney pelvis (Figure 5.21). Some drugs, such as sulphonamide and quinolone antibiotics, precipitate out of solution and cause crystaluria, and often subsequent urinary stone formation. Drugs that affect urinary pH or cause dehydration may cause coadministrated drugs to form urinary crystals (Frazier and Seely, 2013), and the presence of uroliths or 'stones' can cause obstruction of the ureters or urethra, with subsequent damming back of urine to the kidneys and bladder. This lesion is termed 'hydronephrosis' and is characterised by the replacement of kidney tissue with a thin rim of normal renal tissue surrounding a large cystic fluid mass, due to the atrophy of the renal tissue.

Figure 5.21 Urinary stones in the bladder of a mouse.

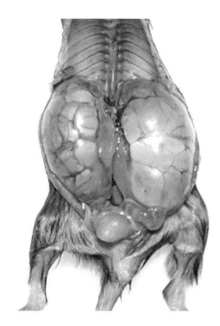

Figure 5.22 Hydronephrosis in both kidneys of a mouse.

Hydronephrosis is also a common congenital lesion in rats and mice, and thus is not always test article-related (Figure 5.22).

Papillary necrosis is a severe lesion that is observed at necropsy and is characterised by a yellow, necrotic medulla or papilla extending into the pelvis of the kidney. Papillary necrosis occurs when the medulla and papilla of the kidney undergo ischaemic necrosis. The most common cause is nonsteroidal anti-inflammatory drugs (NSAIDs), but cyclophosphamide and radiocontrast media may cause similar lesions. Generally, angiotensin-converting enzyme (ACE) inhibitors cause proliferation of the juxtaglomerular apparatus via chronic stimulation of the renin–angiotensin system (Greaves, 1998), but this lesion is only visible under the light microscope. Mineralisation of the kidney may be visualised at necropsy as white cortical surface stippling. Renal mineralisation is a common background lesion, although it may occasionally be a test article-related finding observed through the use of vitamin D analogues. Pigmentation of the kidney is common. Occasionally, lipofuscin accumulation may be associated with compound administration (e.g. benzodiazepines) (Owen et al., 1970). Haemoglobin/iron/haemosiderin may accumulate in the kidneys during drug-induced red-blood-cell breakdown, and myoglobin from muscle degeneration and necrosis can accumulate within kidney tubules. Warfarin (an anti-blood-clotting drug) may cause haematuria (blood in the urine), but so too may the presence of urinary or renal crystals/stones and cystitis (Figure 5.23). Bilirubin may be present in the kidneys as a result of drugs that inhibit bilirubin uptake and transport, and this will cause a yellow-greenish colour on the cortical surface.

Neoplastic changes in the kidney and urinary bladder include benign adenomas and malignant carcinomas, derived from renal tubular or urinary bladder epithelium. Compound or chemical-induced urinary-system tumours have been reported, including N-ethyl-N-hydroxyethylnitrosomide-induced tumours (Konishi et al., 2001). Hyperplastic and neoplastic changes involving renal and urinary bladder epithelium may be caused by cyclophosphamide (Greaves, 1998). Connective-tissue tumours and congenital renal tumours (e.g. nephroblastoma) are also described in the urinary tract. PPAR agonists cause urothelial hypertrophy and hyperplasia, which may produce tumours (Tseng and Tseng, 2012).

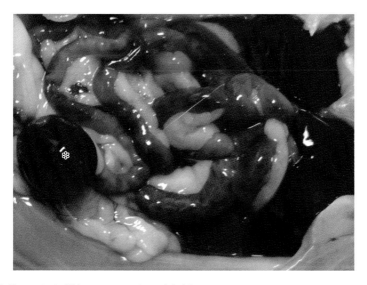

Figure 5.23 Haematuria (*) in a mouse urinary bladder.

5.7 Lymphoreticular System

According to important work on toxicity in the lymphoid system (Haley et al., 2005), potential drug immunotoxicity should not be confirmed unless white blood cell counts, globulin and albumin/globulin ratios, macroscopic and microscopic pathology and thymus and spleen weights have all been evaluated. At necropsy, the pathologist and study director should decide whether the lymphoid organs are hypertrophied or atrophied. In addition, it is useful to establish whether the shape, weight, colour or texture of the lymphoid organs is different in the treated animals. It is important to remember that the thymus is large and visible in young animals and regresses at the time of sexual maturity. The thymus of the minipig pig is situated in the neck region, whilst the thymus of the guinea pig extends into the thoracic cavity.

At necropsy, splenomegaly is characterised by an enlarged spleen (pentobarbitone used as a euthanising agent will cause severe congestion of the spleen in all animal species). Enlarged splenic white pulp is generally caused by hyperplasia of the white pulp or lymphoma and is visible at necropsy (Figure 5.24). Decreased cellularity of the white pulp (i.e. the periarteriolar lymphoid sheaths) is visible upon cutting open the spleen where the white pulp is almost absent. Fibrosis of the capsule is characterised by a thickened wrinkled splenic capsule. Infarction and necrosis of the spleen result in large yellow necrotic wedge-shaped areas visible at necropsy and extending into the cut surface.

At necropsy, the mesenteric and mandibular lymph nodes are often increased in size, as they drain the gastrointestinal tract and the oral cavity, respectively. The mouth is exposed to high numbers of foreign antigens, and the mandibular lymph nodes often react with an increase in follicles, germinal centres, plasma cells (plasmacytosis) and macrophages within the sinuses (sinus histiocytosis), as well as by increasing in size. These changes are generally considered background changes in most rodent and dog studies.

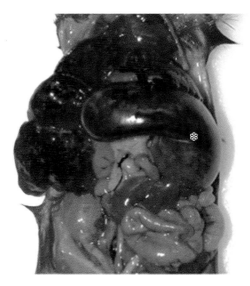

Figure 5.24 Severe enlargement of the spleen (*) due to lymphoma in a mouse.

As rodents, dogs and non-human primates grow older, the lymph nodes undergo atrophy, which is often quite dramatic in rats and mice on carcinogenicity studies. Mucosal-associated lymphoid tissue (MALT) includes NALT, bronchus-associated lymphoid tissue (BALT) (located within the bronchial submucosa of rabbits, rats, mice, dogs and pigs) and gut-associated lymphoid tissue (GALT). MALT is generally found at mucosal surfaces and often demonstrates microscopic lesions, such as a decrease in or hyperplasia of lymphocytes. Tumours (particularly lymphoma) can occur in MALT, GALT and BALT, becoming large and prominent at necropsy.

At necropsy, the bone marrow may display yellow liquefied tissue. This indicates an absence of red-blood-cell precursors and is seen in cases of immunosuppression, such as in anticancer treatments. The bone marrow often displays test article-related findings, because many anticancer drugs (e.g. cyclophosphamide, cyclosporine A, tacrolimus, rapamycin and selective kinase inhibitors) affect the rapidly dividing cells of the bone marrow. Microscopic changes include increases and (especially) decreases in all or just one specific cell subtype, including neutrophils, eosinophils and basophils. Increased numbers of leukocytes is called leukaemia and results in a greater susceptibility to infection. A reduction in red-blood precursors causes anaemia, whilst a reduction in platelets causes thrombocytopaenia with a consequent bleeding tendency. Chloramphenicol suppresses all bone-marrow components, causing a pancytopaenia (Glaister, 1986).

When analysing test article-related effects in the lymphoid system, it is important to distinguish between the effects of stress (which may cause a reduction in lymphocytes) and actual drug-induced effects on the lymphoid organs. Stress effects include decreased body weight, decreased food consumption, decreased thymic weights, increased adrenal weights and a stress leukogram (increased neutrophils and monocytes, with decreased lymphocytes and eosinophils) (Everds et al., 2013).

Among the tumours of the lymphoid system are the lymphomas, which occur commonly and spontaneously in older CD-1, C57BL/6 and B6C3F1 mice and are characterised by large, white, bulging, superficial abdominal and thoracic lymph nodes (Figure 5.25), thymus and spleen; Sprague Dawley rats also develop lymphoma, but not at the same high level as that of mice (Frith, 1988). Other haemopoietic tumours include leukaemias, histiocytic sarcoma, thymoma and mast cell tumours.

5.8 Musculoskeletal System

The most common test article-related injury observed in the muscle is necrosis, which is visible at necropsy as large tan to white areas. Reversible cell change (degeneration) in the skeletal muscle is almost always accompanied by regeneration, which is only visible under the light microscope and consists of bluish new muscle fibres with central nuclei and macrophages phagocytosing the degenerating tissue. A number of xenobiotics can cause myotoxicity: selenium, monensin, PPARs (alpha) and statins (Westwood et al., 2008). Irritant compounds such as tiamulin in pigs cause severe necrosis in the muscles surrounding the area of intramuscular injection.

Enlargement of the myocytes (hypertrophy) is recognised at necropsy by the presence of an increased muscle mass and may be caused by growth factors and growth hormone (Prysor-Jones and Jenkins, 1980). Muscle atrophy is easier to recognise at necropsy than

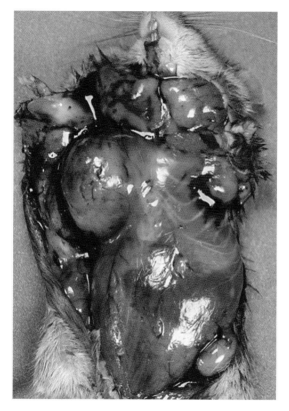

Figure 5.25 Severely enlarged submandibular, axillary and inguinal lymph nodes in lymphoma in a mouse.

under the light microscope. Few compounds cause direct muscle atrophy, but devices that cause immobilisation of limbs or loss of innervation due to peripheral nerve damage, such as hexachlorophene, will cause atrophy of specific muscles (generally those of the limbs). Rhabdomyosarcoma is a malignant tumour of the skeletal muscle and may be induced by certain carcinogens, but this is very rare.

Thickened bones may be observed at necropsy, when they appear harder and more difficult to section. Hyperostosis is defined as an increased amount of bone. This finding may be congenital or may be caused by compounds that control bone formation and resorption (Vahle et al., 2013). Parathyroid hormone may cause hyperostosis (Vahle et al., 2002). Bone atrophy or osteopaenia may be observed with corticosteroid toxicity and pyridoxine deficiency (Vahle et al., 2013); this may only be visible upon histopathological examination of the bone. Necrosis of bone may be difficult to observe at necropsy unless there is a sequestrum formation (a piece of necrotic bone has broken off and become isolated). Bone necrosis may be caused by corticosteroids and biphosphonates (Jones and Allen, 2011).

Fracture and callus formation in long bones (e.g. femur) in laboratory animals may occur due to trauma or may rarely be linked to a compound. Fracture is generally characterised by the presence of bone fragments, haemorrhage and swelling of the area at necropsy. Fibrous osteodystrophy involves concurrent bone resorption by osteoclasts

Figure 5.26 Enlarged parathyroid gland (*) in renal failure in a dog.

and an increase in fibrous tissue infiltration into the medullary cavities. The most common causes of this syndrome are chronic renal failure (occasionally secondary to renal failure caused by a xenobiotic), parathyroid gland hyperplasia (Figure 5.26) and tumour formation and excess vitamin D. Fibrous osteodystrophy may be visible at necropsy, when bones appear rubbery and are easy to bend, although a definitive diagnosis can only be made upon histopathological examination of the bone.

At necropsy, inflammation of the joints is characterised by thickening of the joint capsule and by the presence of ulcerated areas on the articular cartilage. Osteoarthritis (inflammation of the joints) is not generally a test article-related condition, but degeneration of cartilage often is; this is caused by xenobiotics, such as quinolones (Burkhardt et al., 1990). Bone tumours include osteoma and osteosarcoma. Osteosarcoma is an aggressive tumour that is visible at necropsy and may be caused by parathyroid hormone (Vahle et al., 2002). At necropsy, it presents as a large bone swelling, with possible metastasis to the lung. Chondroma and chondrosarcoma are tumours of the cartilage that may be visible at necropsy, whilst synovial sarcoma is a rare tumour of the synovial membrane lining the joint capsule.

5.9 Cardiovascular System

Heart failure in laboratory animals may be difficult to recognise, but often results in a blue tongue and blue mucous members (due to lack of oxygen), rapid breathing and widespread oedema or fluid accumulation. At necropsy, the heart may be enlarged with thin flabby walls. An enlarged heart is easy to recognise at necropsy, and is referred to as 'cardiomyopathy'. The heart responds to a greater workload by increasing myofibre mass (hypertrophy); this may be encountered in trained athletes and in animals and humans treated with alpha and beta blockers and anabolic steroids (Sullivan et al., 1998). Ageing and starvation, as well as ACE inhibitors, are reported to cause atrophy or a decrease in heart weight (Isaacs, 1998).

Abnormal fluid accumulating within the pericardium may be an indication of widespread oedema, whilst blood accumulating in the pericardium (haemopericardium) is often an indication of a bleeding tendency. Cardiovascular disease is common in the Western world, and a number of cardioactive compounds have been developed to treat it, but they may have unwanted side effects. Although the heart has a large physiological capacity, toxic effects in the myocardium have serious effects because cardiomyocytes are unable to regenerate after death by necrosis. This means necrotic myocytes will heal with fibrosis or scar tissue, which may affect heart performance and contraction, largely by causing ventricular dilatation and poor contractibility of the heart muscle.

Developmental abnormalities such as hole in the heart (ventricular septal defect) are easy to see at necropsy and have been linked to phenobarbital and caffeine in rats (Isaacs, 1998). Many drugs have an effect on ion movement, causing irregular conduction (e.g. cardiac glycosides) and leading to severe clinical signs, such as arrhythmia, which may result in death. The heart will appear unremarkable at necropsy: an electrocardiogram before death may be more useful. The heart requires large amounts of oxygen to perform its functions, and thus is vulnerable to the effects of hypoxia (may cause myocyte necrosis), which may be induced by mitochondrial enzyme inhibitors such as cyanide, compounds such as monensin or doxorubicin and vasoactive amines such as noradrenaline. More recent compounds, such as tyrosine kinase inhibitors, cause cardiotoxicity (Chen et al., 2008). Mineralisation is often noted in areas of myocardial necrosis and causes a whitish colour and grittiness in the necrotic area.

At necropsy, it is important to examine the heart valves, particularly in larger animals such as dogs and non-human primates. Haemorrhage, inflammation and fibrosis of the atrioventricular valves of dogs are seen after treatment with positive inotropic agents and vasodilation agents (such as hydralazine), which cause turbulent blood flow (Isaacs, 1998). Appetite suppressants such as fenfluramine also cause thickened valves in humans (endocardiosis) (Connolly et al., 1997). Underlying heart damage and turbulent blood flow may result in valvular thrombi, which are often associated with indwelling cardiac catheters and may become infected, causing valvular vegetative endocarditis. Lipofuscin pigmentation of the heart (causing a brown colour) may be seen with old age and can occur with chloroquine treatment. Cardiac tumours are rare in laboratory animals, although Schwann cell tumours are encountered in rats spontaneously and as a result of methylnitrosourea (MNU) treatment. Cardiomyopathy is a background lesion noted in older rats and mice (see Chapter 4).

The vascular system is governed by the factors affecting laminar flow, turbulence and viscosity of the blood. It is not easy to recognise changes in the vascular system at necropsy, but study personnel may notice haemorrhage and necrosis surrounding the jugular vein where an intravenous drug was administrated, indicating vascular irritation or poor venous injection technique. Oedema is easy to recognise at necropsy (see Chapter 3). Limbs with dependent oedema have watery, gelatinous subcutaneous tissue and pit upon pressure on the skin and clear fluid is present in the thoracic and abdominal cavities, as well as in the pericardium. Conditions such as hypertension, lymphatic blockage and hypoproteinaemia will result in fluid accumulation (oedema) of tissues. Thus, drugs causing hypoproteinaemia (e.g. by causing renal failure and proteinuria), as well as others, such as mannitol (increased vascular permeability), will cause widespread oedema. Vasoactive drugs such as ergot alkaloids cause vasoconstriction, and thus possible dry gangrene of extremities. Agents that interfere with

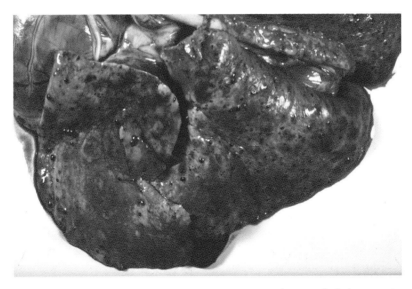

Figure 5.27 Small red raised areas in the lung of a dog, indicating the spread of a haemangiosarcoma from the spleen.

platelet function or blood clotting (such as heparin) have the capacity to cause widespread haemorrhage.

Inflammation of the blood vessels is referred to as 'arteritis' or 'vasculitis' and is not easy to recognise at necropsy unless the damage is extremely severe (e.g. perivascular jugular vein inflammation due to an injected irritant drug). Vasculitis may be induced by drugs such as sulphonamides (Isaacs, 1998). Thickening of blood vessels includes proliferation of various layers of the vessel wall; this may be observed in dogs treated with phosphodiesterase (PDE) III inhibitors and drugs that cause hypertension (Louden and Brott, 2013). Medial necrosis of the smooth muscle of blood vessels can be caused by vasoactive agents such as minoxidil (Isaacs, 1998). These conditions are only visible under the light microscope.

Tumours of the vascular system are rare, although haemangioma and haemangiosarcoma (Figure 5.27) can be induced by PPARs (Hardisty et al., 2007) and are easy to recognise at necropsy, as they have a characteristic blood-red colour. Repeated intravascular administration of compounds can result in the formation of hair and skin fragments, which lodge in the lung and induce a granulomatous reaction around the foreign material. Recently, a number of novel plasma biomarkers (e.g. troponin) have enhanced the diagnosis (noninvasive) and prognosis of adverse cardiac events in both laboratory animals and humans. In addition, biochemical biomarkers of vascular toxicity (e.g. von Willebrand factor, endothelin; Louden and Brott, 2013) have also been described.

5.10 Endocrine System

The endocrine glands (pituitary, adrenals, thyroid, parathyroid and pancreatic islets) work together to provide feedback loops to various organs, providing control of

Figure 5.28 Pituitary adenoma (large, dark red mass at the base of the cranial cavity) (*) in a rat.

physiological processes such as the release of adrenalin from the adrenals enabling the fight-or-flight response.

Pituitary enlargement is often seen at necropsy in ageing rats and mice, particularly female rats, causing compression of the surrounding normal brain tissue (Figure 5.28). Many of the pituitary tumour cells are prolactin-producing, which can lead to an increase in mammary tumours in rats and mice. Pituitary carcinoma is a more invasive tumour, but is quite rare in rats and mice. Ovariectomy, castration and the administration of the oral contraceptive pill produce a decrease of oestrogens and testosterone, causing an enlargement of the pituitary as it attempts to produce more follicle stimulating hormone (FSH) and luteinising hormone (LH) in order to stimulate more oestrogen production.

Enlargement of the thyroids can sometimes be seen at necropsy (Figure 5.29). Hyperplasia and hypertrophy occur in response to chronic excessive TSH stimulation, generally because T3 and T4 levels are low. This is seen in cases of increased clearance of thyroid hormone, which occurs secondary to hepatocellular microsomal enzyme induction (e.g. with phenobarbital, benzodiazepine and steroid treatment). Many compounds induce thyroid hyperplasia, including carbimazole and lithium. The study director will be aware of the increased size and weight of the liver at necropsy, prompting an examination of the thyroids (larger) and pituitary (larger). Enlarged thyroids may be difficult to recognise at necropsy, particularly in rodents; however, an increase in thyroid weights is a good indicator of test article-induced thyroid hyperplasia. Black discolouration of the thyroid is seen with minocycline administration and is easy to recognise at necropsy. Tumours of the thyroid include follicular adenomas and follicular carcinomas.

Changes in the parathyroid gland are virtually impossible to visualise in rodents at necropsy, but in larger animals it may be possible to see the enlarged parathyroid gland

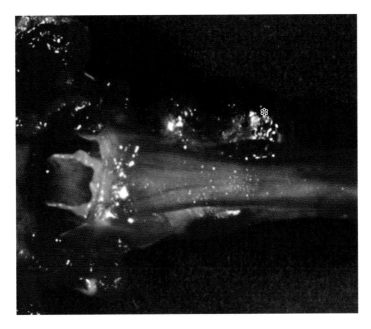

Figure 5.29 Enlargement of the thyroid (*) in a rat.

within the thyroid glands. Compounds that alter calcium uptake can cause proliferative lesions in the chief cells responsible for producing parathyroid hormone in the parathyroid gland. Proliferative changes in the parathyroid glands are commonly caused by renal failure (particularly in old male rats), low-calcium diets, irradiation and steroid and calcitonin hormone treatment.

An increase in adrenal weight and size can be observed at necropsy. The adrenal gland is vulnerable to toxicity by xenobiotics, due to its high lipid content and high number of blood vessels. Proliferative lesions in the adrenal cortex include hyperplasia and hypertrophy induced by adrenocorticotropic hormone (ACTH). Atrophy of the adrenal occurs as a result of a deficiency of ACTH, generally due to corticosteroid administration. Vacuolation (caused by ketoconazole and triaryl, amongst other compounds) and necrosis are common test article-related lesions in the adrenal cortex, but are generally only visible under the light microscope. Masses in the adrenal include adenoma and carcinoma, which are relatively rare in rats and mice. The adrenal medulla is seldom affected by compounds, but phaeochromocytomas (tumour of the medulla; i.e. chromataffin cells) (Figure 5.30) are common in older male rats. Compounds that cause adrenal medullary proliferation and phaeochromocytomas include reserpine and vitamin D3 (Rosol et al., 2001).

Lesions of the pancreatic islets can only be visualised under the light microscope. They include vacuolation (caused by streptozotocin) and amyloid deposition (seen in non-human primates). Alloxan and stroptozotocin cause beta-cell necrosis and are thus often used to create animal models of diabetes mellitus. Cyclosporin A and zinc chelators (Taylor, 2005) also cause beta-cell lesions. Proliferative masses in the islets may be visible at necropsy, and include islet-cell adenoma and islet-cell carcinomas. Heliotrine may cause beta-cell adenomas in rats (Chandra et al., 2013).

Figure 5.30 Phaemochromocytoma in an adrenal adjacent to the kidney in a rat.

5.11 Reproductive System

Test article-related findings in the reproductive system generally herald the end of a particular compound, as human fertility is considered too fragile to compromise. Maturity of the reproductive tissues is important in study design, since if a study is conducted in which animals are still sexually immature at the end of dosing, then the possibility exists that the drug may still be a reproductive toxicant. This is a greater problem in dogs and non-human primates than in rodents, which mature at about 10 weeks of age. Organ weights (particularly of the testes and epididymis), sperm analysis, endocrine measurements and reproductive cyclicity data can be more sensitive than histopathology. Prostate and seminal vesicle weights are also important in assessing drug-related toxicity and may be more sensitive than histopathological examination.

Most testicular changes within the seminiferous tubules are only visible under the light microscope, although severe testicular atrophy will result in small testes at necropsy

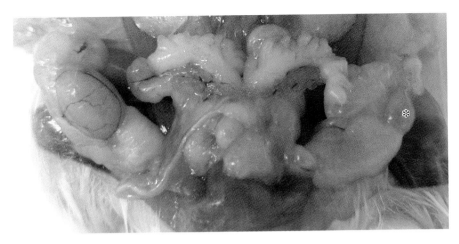

Figure 5.31 Testicular atrophy (*) in the left testis of a mouse.

(Figure 5.31). Germ-cell degeneration is a common drug-induced change seen with androgen deficiency (Troiano et al., 1994) and cytotoxic anticancer drugs. Compound-induced changes also include lesions within the Sertoli cells that provide support to the spermatogonia.

Proliferative changes and tumours in the Leydig cells are common in older rats, but are not visible at necropsy and are not generally thought to be relevant to humans. Adenomas and carcinomas of the rete testis can occasionally be associated with compounds, but in general tumours such as Sertoli-cell tumour, seminoma and teratomas are rare and unrelated to compound administration.

Ovarian atrophy is recognised at necropsy by the presence of small, hard ovaries. Selective oestrogen-receptor modulators (SERMs) such as tamoxifen cause ovarian atrophy (Rehm et al., 2007a). Ovarian cysts (Figure 5.32) are easy to observe at necropsy. They may sometimes be quite large and filled with clear fluid or blood. Ovarian cysts increase as rodents get older, but they may be treatment-related in the case of LH disruption, increased oestrogen production and increased androgen levels (Vidal et al., 2013).

Increases and decreases in the corpora lutea can be observed at necropsy by counting the small raised masses present on the surface of the ovary. Increased corpora lutea is seen with bromocriptine treatment, due to the inhibition of prolactin (Rehm et al., 2007b), and enlarged corpora lutea are observed with an increase in ovarian weight following angiogenesis-inhibitor treatment. Tumours of the ovary are unlikely to be test article-related, although mesovarial leiomyomas are associated with beta-receptor agonists (Gopinath and Gibson, 1987).

Macroscopic changes in the uterus are rare at necropsy, but cystic endometrial hyperplasia is recognisable by the presence of thickened, tortuous uterine tubes. Here, the uterus is enlarged with dilated endometrial glands. This is seen in older mice and rats. In dogs, cystic endometrial hyperplasia often occurs in conjunction with pyometra (pus-filled uterus) and is associated with progesterone treatment.

Proliferative changes in the uterus include endometrial polyp (particularly in older rats and mice), endometrial adenoma and adenocarcinoma and leiomyoma, which may be induced by long-term use of oestrogenic compounds. Excessive lactation is seen at

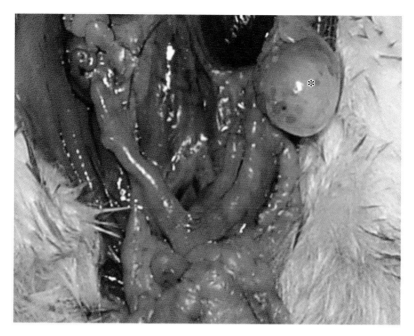

Figure 5.32 Ovarian cyst (*) in a mouse.

necropsy in rats and mice and is linked to pregnancy or treatment with progesterone and oestrogens. In male rats, oestrogenic compounds cause the mammary tissue to resemble that of female rats (histologically), whilst female rats undergo male mammary gland differentiation with androgen treatment. Hyperplasia, adenoma, carcinoma and fibroadenoma (particularly of the older rat) (Figure 5.33) are common mammary-gland tumours, and are generally not test article-related.

5.12 Central and Peripheral Nervous System

Public concern about the possibility of drugs causing neurological side effects has meant that the guidelines for assessing neurotoxicity in animal studies are very stringent. The central nervous system is very susceptible to toxic damage, due to its high dependency on glucose and oxygen, its high lipid content (which makes it easy for lipid-soluble compounds (such as anaesthetic agents) to cross the blood–brain barrier and accumulate in the brain) and the fact that neurons cannot be replaced if damaged. Most of the sensory and motor-related pathways pass to the thalamus, which processes this information. The limbic system (olfactory lobes, hippocampus and connections to the hypothalamus) is responsible for emotion, memory and reproductive behaviours. The hypothalamus is responsible for endocrine functions, the cerebellum is concerned with balance and the medulla oblongata controls the life-maintaining centres, such as respiration, heart rate and blood pressure. Sections from each of these parts of the brain, as well as the peripheral nerves (sciatic), are recommended for histopathological examination. Resin-embedding, followed by staining with toluidine blue, gives excellent morphology of peripheral nerves.

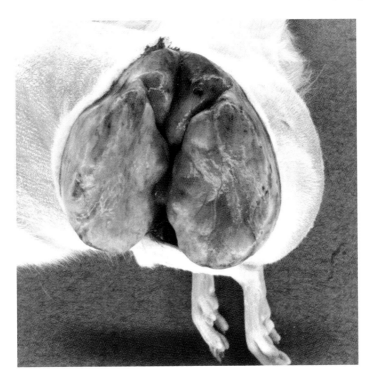

Figure 5.33 Mammary gland fibroadenoma in a rat.

The clinical signs displayed by animals during a study give important clues about where a test article-related lesion may be located in the central or peripheral nervous system. The increased tight junctions in the endothelial cells of the brain blood vessels and the reduced pores and pinocytosis make up the blood–brain barrier, which provides some protection against internal and external brain toxins present in the blood. Most changes in the central nervous system are identified under the light microscope and not at necropsy. Lesions include neuronal necrosis, neuronal vacuolation (the pathologist should be certain this is not an artefact), nerve-fibre degeneration, and gliosis (accumulation of glial cells around an area of injury). Vacuolation in the cytoplasm or nucleus of neurons or within the myelin sheath is often observed; this is generally an artefact caused by processing, as suggested by the absence of any inflammatory cells and the lack of bilateral involvement of paired structures in the brain. Infarction is seen with treatments that restrict or occlude blood flow to the brain. Gitter cells are macrophages containing lipid and are seen in areas of injury.

Hypoxia due to various mechanisms causes neuron necrosis. At necropsy, tissues such as the mucous membranes may be blue (lack of red oxygenated blood), whilst the blood may be brown (nitrate toxicity) or cherry red (carbon monoxide poisoning). Chromatolysis is a microscopic lesion characterised by eccentric nuclei and margination of the Nissl substance within neurons. It is caused by compounds such as acrylamide. Convulsions during the in-life period are often a clear indication of central nervous system toxicity. Strychnine causes tonic convulsions in response to loud noises or bright lights. Paralysis or dragging of limbs suggests nerve damage. Triethyl tin and hexachlorophene both cause myelin damage (myelinopathy) (Butt et al., 2013).

Purulent material (greenish, yellow pus) present on the surface of the brain is seen at necropsy in cases of meningitis. This lesion is unusual in toxicity and safety studies, but it may occur if the compound causes severe immunosuppression and hence infection of the meninges by bacteria such as *Streptococcus* sp. Indwelling cerebral or ventricular catheters can cause similar lesions by introducing bacteria into the sterile brain environment. A head tilt can indicate a lesion in the cerebrum, and incoordination and circling can indicate a cerebellar lesion with tremors. Brain oedema and swelling lead to coma and loss of consciousness. Heavy metals such as lead and cadmium can cause endothelial swelling with an increase in vascular leakiness, which results in cerebral oedema.

It is difficult to see tumours in the brain at necropsy, as they tend to be embedded within the brain tissue. Proliferative lesions are rare in the central nervous system, but include astrocytomas, oligodendrogliomas, granular cell tumours and schwannomas. Malignant schwannomas can be induced by 1,1-dimethyl-hydrazine treatment. Acrylonitrile and ethylene oxide are neurocarcinogens that cause tumours of the central nervous system (Buckley, 1998).

5.13   Ear

In the inner ear, cells such as hair cells in the cochlea and the vestibular epithelium can be destroyed by drugs such as aminoglycoside antibiotics (Imamura and Adams, 2003), quinine, salicyclates and cisplatin. This change can only be confirmed using light microscopy of the inner ear.

References

Aguirre, S.A., Huang, W., Prasanna, G. and Jessen, B. (2009) Corneal neovascularization and ocular irritancy responses in dogs following topical ocular administration of an EP4-prostaglandin E2 agonist. *Toxicologic Pathology*, 37, 911–20.

Betton, G.R. (1998a) The digestive system. I: The gastrointestinal tract and exocrine pancreas. In: Turton, J. and Hoosen, J. (eds). Target Organ Pathology: A Basic Text, Taylor & Francis, London, pp. 29–60.

Betton, G.R. (1998b) The digestive system. II: The hepatobilliary system. In: Turton, J. and Hoosen, J. (eds). Target Organ Pathology: A Basic Text, Taylor & Francis, London, pp. 61–97.

Buckley, P. (1998) The nervous system. In: Turton, J. and Hoosen, J. (eds). Target Organ Pathology: A Basic Text, Taylor & Francis, London, pp. 273–310

Burkhardt, J.E., Hill, M.A., Carlton, W.W. and Kesterson, J.W. (1990) Histologic and histochemical changes in articular cartilages of immature beagle dogs dosed with difloxacin, a fluoroquinolone. *Veterinary Pathology*, 27, 162–70.

Butt, M.T., Sills, R. and Bradley, A. (2013) Nervous system. In: Sahota, P.S., Popp, J.A., Hardisty, J.F. and Gopinath, C. (eds). Toxicologic Pathology: Nonclinical Safety Assessment, CRC Press, Boca Raton, FL, pp. 896–930.

Cattley, R.C., Popp, J.A. and Vonderfecht, S.L. (2013) Liver, gallbladder, and exocrine pancreas. In: Sahota, P.S., Popp, J.A., Hardisty, J.F. and Gopinath, C. (eds). Toxicologic Pathology: Nonclinical Safety Assessment, CRC Press, Boca Raton, FL, pp. 367–421.

Chandra, S.A., Nolan, M.W. and Malarkey, D.E. (2010) Chemical carcinogenesis of the gastrointestinal tract in rodents: an overview with emphasis on NTP carcinogenesis bioassays. *Toxicologic Pathology*, 38, 188–97.

Chandra, S., Hoenerhoff, M.J. and Peterson, R. (2013) Endocrine glands. In: Sahota, P.S., Popp, J.A., Hardisty, J.F. and Gopinath, C. (eds). Toxicologic Pathology: Nonclinical Safety Assessment, CRC Press, Boca Raton, FL, pp. 655–716.

Chen, M.H., Kerkelä, R. and Force, T. (2008) Mechanisms of cardiac dysfunction associated with tyrosine kinase inhibitor cancer therapeutics. *Circulation*, 118, 84–95.

Clarke, J., Hurst, C., Martin, P., Vahle, J., Ponce, R., Mounho, B., Heidel, S., Andrews, L., Reynolds, T. and Cavagnaro, J. (2008) Duration of chronic toxicity studies for biotechnology-derived pharmaceuticals: is 6 months still appropriate? *Regulatory Toxicology and Pharmacology*, 25, 130–45.

Connolly, H.M., Crary, J.L., McGoon, M.D., Hensrud, D.D., Edwards, B.S., Edwards, W.D. and Schaff, H.V. (1997) Valvular heart disease associated with fenfluramine-phentermine. *New England Journal of Medicine*, 337, 581–8.

Contrera, J.F., Jacobs, A.C. and Degeorge, J.J. (1997) Carcinogenicity testing and the evaluation of regulatory requirements for pharmaceuticals. *Regulatory Toxicology and Pharmacology*, 50, 2–22.

Diaz, D., Allamneni, K., Tarrant, J.M., Lewin-Koh, S.C., Pai, R., Dhawan, P., Cain, G.R., Kozlowski, C., Hiraragi, H., La, N., Hartley, D.P., Ding, X., Dean, B.J., Bheddah, S. and Dambach, D.M. (2012) Phosphorous dysregulation induced by MEK small molecule inhibitors in the rat involves blockade of FGF-23 signaling in the kidney. *Toxicological Sciences*, 125, 187–95.

Donald, S., Verschoyle, R.D., Edwards, R., Judah, D.J., Davies, R., Riley, J., Dinsdale, D., Lopez Lazaro, L., Smith, A.G., Gant, T.W., Greaves, P. and Gescher, A.J. (2002) Hepatobiliary damage and changes in hepatic gene expression caused by the antitumor drug ecteinascidin-743 (ET-743) in the female rat. *Cancer Research*, 62, 4256–62.

Elcock, L.E., Stuart, B.P., Wahle, B.S., Hoss, H.E., Crabb, K., Millard, D.M., Mueller, R.E., Hastings, T.F. and Lake, S.G. (2001) Tumors in long-term rat studies associated with microchip animal identification devices. *Experimental and Toxicologic Pathology*, 52, 483–91.

Ettlin, R.A., Kuroda, J., Plassmann, S. and Prentice, D.E. (2010) Successful drug development despite adverse preclinical findings part 1: processes to address issues and most important findings. *Journal of Toxicologic Pathology*, 23, 189–211.

Everds, N.E., Snyder, P.W., Bailey, K.L., Bolon, B., Creasy, D.M., Foley, G.L., Rosol, T.J. and Sellers, T. (2013) Interpreting stress responses during routine toxicity studies: a review of the biology, impact, and assessment. *Toxicologic Pathology*, 41, 560–614.

Ferguson, J. (2002) Photosensitivity due to drugs. *Photodermatology, Photoimmunology & Photomedicine*, 18, 262–9.

Frazier, K.S. and Seely, J.C. 2013. Urinary system. In: Sahota, P.S., Popp, J.A., Hardisty, J.F. and Gopinath, C. (eds). Toxicologic Pathology: Nonclinical Safety Assessment, CRC Press, Boca Raton, FL, pp. 421–84.

Frazier, K.S., Seely, J.C., Hard, G.C., Betton, G., Burnett, R., Nakatsuji, S., Nishikawa, A., Durchfeld-Meyer, B. and Bube, A. (2012) Proliferative and nonproliferative lesions of the rat and mouse urinary system. *Toxicologic Pathology*, 40(4 Suppl.), 14S–86S.

Frith, C.H. (1988) Morphologic classification and incidence of hematopoietic neoplasms in the Sprague-Dawley rat. *Toxicologic Pathology*, 16, 451–7.

Glaister, J.R. (1986) Principles of Toxicological Pathology, Taylor & Francis, London.

Gopinath, C. and Gibson, W.A. (1987) Mesovarian leiomyomas in the rat. *Environmental Health Perspectives*, 73, 107–13.

Grattagliano, I., Bonfrate, L., Diogo, C.V., Wang, H.H., Wang, D.Q. and Portincasa, P. (2009) Biochemical mechanisms in drug-induced liver injury: certainties and doubts. *World Journal of Gastroenterology*, 15, 4865–76.

Greaves, P. (1998) The urinary system. In: Turton, J. and Hoosen, J. (eds). Target Organ Pathology: A Basic Text, Taylor & Francis, London, pp. 89–126.

Greene, G.W. Jr, Collins, D.A. and Bernier, J.L. (1960) Response of embryonal odontogenic epithelium in the lower incisor of the mouse to 3-methylcholanthrene. *Archives of Oral Biology*, 1, 325–32.

Haley, P.J. (2003) Species differences in the structure and function of the immune system. *Toxicology*, 188, 49–71.

Haley, P., Perry, R., Ennulat, D., Frame, S., Johnson, C., Lapointe, J.M., Nyska, A., Snyder, P., Walker, D. and Walter, G.; STP Immunotoxicology Working Group. (2005) STP position paper: best practice guideline for the routine pathology evaluation of the immune system. *Toxicologic Pathology*, 33, 404–7.

Hard, G.C., Alden, C.L. and Bruner, R.H., 1999. Non-proliferative lesions of the kidney and lower urinary tract in rats. In: URG-1 Guides for Toxicologic Pathology, STP/ARP/AFIP, Washington, DC.

Hardisty, J.F., Elwell, M.R., Ernst, H., Greaves, P., Kolenda-Roberts, H., Malarkey, D.E., Mann, P.C. and Tellier, P.A. (2007) Histopathology of hemangiosarcomas in mice and hamsters and liposarcomas/fibrosarcomas in rats associated with PPAR agonists. *Toxicologic Pathology*, 35, 928–41.

Harkema, J.R. (1991) Comparative aspects of nasal airway anatomy: relevance to inhalation toxicology. *Toxicologic Pathology*, 19, 321–36.

Harkness, J.E. and Ridgway, M.D. (1980) Chromodacryorrhea in laboratory rats (Rattus norvegicus): etiologic considerations. *Laboratory Animal Science*, 30, 841–4.

Haschek, W.M., Rousseaux, C.G. and Wallig, M.A. (2010) Skin and oral mucosa. In: Haschek, W.M., Rousseaux, C.G. and Wallig, M.A. (eds). Fundamentals of Toxicologic Pathology, 2nd edn, Academic Press, New York, pp. 135–61.

Hirano, T., Manabe, T., Ando, K. and Tobe, T. (1992) Acute cytotoxic effect of cyclosporin A on pancreatic acinar cells in rats. Protective effect of the synthetic protease inhibitor E3123. *Scandinavian Journal of Gastroenterology*, 27, 103–7.

Imamura, S. and Adams, J.C. (2003) Changes in cytochemistry of sensory and nonsensory cells in gentamicin-treated cochleas. *Journal of the Association for Research in Otolaryngology*, 4, 196–218.

Ingram, A.J. (1998) The Inteumentary System. In: Turton, J. and Hoosen, J. (eds). Target Organ Pathology: A Basic Text, Taylor & Francis, London, pp. 1–28.

Isaacs, K.R. (1998) The cardiovascular system. In: Turton, J. and Hoosen, J. (eds). Target Organ Pathology: A Basic Text, Taylor & Francis, London, pp. 141–76.

Jones, L. and Allan, M. (2011) Animal models of osteonecrosis. *Clinical Reviews in Bone and Mineral Metabolism*, 9, 63–80.

Konishi, N., Nakamura, M., Ishida, E., Kawada, Y., Nishimine, M., Nagai, H. and Emi, M. (2001) Specific genomic alterations in rat renal cell carcinomas induced by N-ethyl-N-hydroxyethylnitrosamine. *Toxicologic Pathology*, 29, 232–6.

Lewis, D.J. and McKevitt, T.P. (2013) Respiratory system. In: Sahota, P.S., Popp, J.A., Hardisty, J.F. and Gopinath, C. (eds). Toxicologic Pathology: Nonclinical Safety Assessment, CRC Press, Boca Raton, FL, pp. 367–421.

Lewis, D.J., Williams, T.C. and Beck, S.L. (2014) Foamy macrophage responses in the rat lung following exposure to inhaled pharmaceuticals: a simple, pragmatic approach for inhaled drug development. *Journal of Applied Toxicology*, 34, 319–31.

Linton, A.L., Clark, W.F., Driedger, A.A., Turnbull, D.I. and Lindsay, R.M. (1980) Acute interstitial nephritis due to drugs: review of the literature with a report of nine cases. *Annals of Internal Medicine*, 93, 735–41.

Louden, C. and Brott, D. (2013) Cardiovascular system. In: Sahota, P.S., Popp, J.A., Hardisty, J.F. and Gopinath, C. (eds). Toxicologic Pathology: Nonclinical Safety Assessment, CRC Press, Boca Raton, FL, pp. 589–654.

Markovits, J.E., Betton, G.R., McMartin, D.N. and Turner, O.C. (2013) Gastrointestinal tract. In: Sahota, P.S., Popp, J.A., Hardisty, J.F. and Gopinath, C. (eds). Toxicologic Pathology: Nonclinical Safety Assessment, CRC Press, Boca Raton, FL, pp. 257–313.

Oesch, F., Fabian, E., Oesch-Bartlomowicz, B., Werner, C. and Landsiedel, R. (2007) Drug-metabolizing enzymes in the skin of man, rat, and pig. *Drug Metabolism Reviews*, 39, 659–98.

Owen, G., Smith, T.H. and Agersborg, H.P. Jr. (1970) Toxicity of some benzodiazepine compounds with CNS activity. *Toxicology and Applied Pharmacology*, 16, 556–70.

Patyna, S., Arrigoni, C., Terron, A., Kim, T.W., Heward, J.K., Vonderfecht, S.L., Denlinger, R., Turnquist, S.E. and Evering, W. (2008) Nonclinical safety evaluation of sunitinib: a potent inhibitor of VEGF, PDGF, KIT, FLT3, and RET receptors. *Toxicologic Pathology*, 36, 905–16.

Prysor-Jones, R.A. and Jenkins, J.S. (1980) Effect of excessive secretion of growth hormone on tissues of the rat, with particular reference to the heart and skeletal muscle. *Journal of Endocrinology*, 85, 75–82.

Reasor, M.J., Hastings, K.L. and Ulrich, R.G. (2006) Drug-induced phospholipidosis: issues and future directions. *Expert Opinions in Drug Safety*, 5, 567–83.

Rehm, S., Stanislaus, D.J. and Williams, A.M. (2007a) Estrous cycle-dependent histology and review of sex steroid receptor expression in dog reproductive tissues and mammary gland and associated hormone levels. *Birth Defects Research. Part B, Developmental and Reproductive Toxicology*, 80, 233–45.

Rehm, S., Stanislaus, D.J. and Wier, P.J. (2007b) Identification of drug-induced hyper- or hypoprolactinemia in the female rat based on general and reproductive toxicity study parameters. *Birth Defects Research. Part B, Developmental and Reproductive Toxicology*, 80, 253–7.

Remuzzi, G. and Perico, N. (1995) Cyclosporine-induced renal dysfunction in experimental animals and humans. *Kidney International. Supplement*, 52, S70–4.

Rosol, T.J., Yarrington, J.T., Latendresse, J. and Capen, C.C. (2001) Adrenal gland: structure, function, and mechanisms of toxicity. *Toxicologic Pathology*, 29, 41–8.

Roth, F.D., Quintar, A.A., Echevarria, E.M.U., Torres, A.I., Aoki, A. and Malonaldo, C.A. (2007) Budesonide effects on Clara cell under normal and allergic inflammatory conditions. *Histochemistry and Cell Biology*, 127, 55–68.

Samberg, M.E., Oldenburg, S.J. and Monteiro-Riviere, N.A. (2010) Evaluation of silver nanoparticle toxicity in skin in vivo and keratinocytes in vitro. *Environmental Health Perspectives*, 118, 407–13.

Strocchi, P., Dozza, B., Pecorella, I., Fresina, M., Campos, E. and Stirpe, F. (2005) Lesions caused by ricin applied to rabbit eyes. *Investigative Ophthalmology & Visual Science*, 46, 1113–16.

Suarez, C.J., Dintzis, S.M. and Frevert, C.W. (2012) Respiratory. In: Treuting, P.M. and Dintzis, S. (eds). Comparative Anatomy and Histology: A Mouse and Human Atlas, Elsevier, Amsterdam, pp. 121–34.

Sullivan, M.L., Martinez, C.M., Gennis, P. and Gallagher, E.J. (1998) The cardiac toxicity of anabolic steroids. *Progress in Cardiovascular Diseases*, 41, 1–15.

Tadokoro, T., Rouzaud, F., Itami, S., Hearing, V.J. and Yoshikawa, K. (2003) The inhibitory effect of androgen and sex-hormone-binding globulin on the intracellular cAMP level and tyrosinase activity of normal human melanocytes. *Pigment Cell Research*, 16, 190–7.

Taylor, C.G. (2005) Zinc, the pancreas, and diabetes: insights from rodent studies and future directions. *Biometals*, 18, 305–12.

Ten Hagen, K.G., Balys, M.M., Tabak, L.A. and Melvin, J.E. (2002) Analysis of isoproterenol-induced changes in parotid gland gene expression. *Physiological Genomics*, 8, 107–14.

Thoolen, B., Ten Kate, F.J., van Diest, P.J., Malarkey, D.E., Elmore, S.A. and Maronpot, R.R. (2012) Comparative histomorphological review of rat and human hepatocellular proliferative lesions. *Journal of Toxicologic Pathology*, 25, 189–99.

Troiano, L., Fustini, M.F., Lovato, E., Frasoldati, A., Malorni, W., Capri, M., Grassilli, E., Marrama, P. and Franceschi, C. (1994) Apoptosis and spermatogenesis: evidence from an in vivo model of testosterone withdrawal in the adult rat. *Biochemical and Biophysical Research Communications*, 202, 1315–21.

Tseng, C.H. and Tseng, F.H. (2012) Peroxisome proliferator-activated receptor agonists and bladder cancer: lessons from animal studies. *Journal of Environmental Science and Health. Part C, Environmental Carcinogenesis & Ecotoxicology Reviews*, 30, 368–402.

Vahle, J.L., Sato, M., Long, G.G., Young, J.K., Francis, P.C., Engelhardt, J.A., Westmore, M.S., Linda, Y. and Nold, J.B. (2002) Skeletal changes in rats given daily subcutaneous injections of recombinant human parathyroid hormone (1-34) for 2 years and relevance to human safety. *Toxicologic Pathology*, 30, 312–21.

Vahle, J.L., Leininger, J.R., Long, P.H., Hall, D.G. and Ernst, H. (2013) Bone, muscle and tooth. In: Sahota, P.S., Popp, J.A., Hardisty, J.F. and Gopinath, C. (eds). Toxicologic Pathology: Nonclinical Safety Assessment, CRC Press, Boca Raton, FL, pp. 561–88.

Vidal, J.D., Mirsky, M.L., Colman, K., Whitney, K.M. and Creasy, D.M. (2013) Reproductive system and mammary gland. In: Sahota, P.S., Popp, J.A., Hardisty, J.F. and Gopinath, C. (eds). Toxicologic Pathology: Nonclinical Safety Assessment, CRC Press, Boca Raton, FL, pp. 717–830.

Westwood, F.R., Duffy, P.A., Malpass, D.A., Jones, H.B. and Topham, J.C. (1995) Disturbance of macrophage and monocyte function in the dog by a thromboxane receptor antagonist: ICI 185,282. *Toxicologic Pathology*, 23, 373–84.

Westwood, F.R., Scott, R.C., Marsden, A.M., Bigley, A. and Randall, K. (2008) Rosuvastatin: characterization of induced myopathy in the rat. *Toxicologic Pathology*, 36, 345–52.

Wilson, D.V. and Walshaw, R. (2004) Postanesthetic esophageal dysfunction in 13 dogs. *Journal of the American Animal Hospital Association*, 40(6), 455–60.

Wojcinski, Z.W., Andrews-Jones, L., Aibo, D.I. and Dunstan, R. (2013) Skin. In: Sahota, P.S., Popp, J.A., Hardisty, J.F. and Gopinath, C. (eds). Toxicologic Pathology: Nonclinical Safety Assessment, CRC Press, Boca Raton, FL, pp. 831–94.

Xiao, K., Li, Y., Luo, J., Lee, J.S., Xiao, W., Gonik, A.M., Agarwal, R.G. and Lam, K.S. (2011) The effect of surface charge on in vivo biodistribution of PEG-oligocholic acid based micellar nanoparticles. *Biomaterials*, 32, 3435–46.

Young, J.T. (1981) Histopathological examination of the rat nasal cavity. *Fundamental and Applied Toxicology*, 1, 309–12.

6

Clinical Pathology

Barbara von Beust

Independent consultant, Winterthur, Switzerland

Learning Objectives

- Understand haematology.
- Understand clinical chemistry.
- Understand urinalysis.
- Understand biomarkers and organ weights.
- Understand clinical pathology profiles in toxicologic pathology.

Clinical pathology is the study of diseases in animals by examination of blood, tissues and fluids; in the context of preclinical safety, clinical pathology focuses primarily on in-life variables, in contrast to anatomic pathology, which is the gold standard for end-point assessment. The most common tissue to use for the in-life assessment of test article-related effects is blood and its components, and this chapter will thus explain in detail the characteristics of this relatively easily accessible and dynamically changing, life-supporting fluid in traditional clinical pathology. Other body fluids, such as urine, cerebrospinal fluid and bronchoalveolar lavage specimens can also be subject to clinical pathologic investigation, which in general measures analytes and other variables (Tomlinson et al., 2013). A compilation of nontraditional clinical pathology applications can be found in Jordan et al. (2014).

Other in-life variables that are important for the assessment of test article-related effects are very basic, and include food consumption and body weight gain or loss. Organ weights can also provide valuable information about the effect of a compound on a tissue at the time of the study end point.

6.1 Clinical Pathology in Study Phases and Good Laboratory Practice

The collection of clinical pathology data represents an integral part of a study. It requires not only careful planning (including the scheduling of blood sampling time points and the provision of recommended volumes and variables), but also a professional and competent laboratory in which to perform the analysis, including quality control and assurance. The evaluation of the resulting data is best performed by experts trained in all aspects of clinical pathology, and in general, it is recommended that study personnel consult or appoint a specialist veterinary clinical pathologist for this purpose (Tomlinson et al., 2013).

6.1.1 Preanalytic Phase: Study Plan

The preanalytic phase includes the overall planning of a particular study, the determination of the animal species, sex, number of animals and schedule of blood samplings, as well as the list of variables to be measured and the equipment to be used (Hall and Everds, 2003; Jordan et al., 2014; Braun et al., 2015). All aspects of the preanalytic phase may influence the validity and scientific meaning of the collected data (Gunn-Christie et al., 2012; Vap et al., 2012). For instance, excessive blood sampling from small species such as mice, rats and even monkeys will result in a compensatory response from the bone marrow to replace the removed blood volume. This can affect the statistical relevance of data analyses (e.g. standard haematology variables such as red blood cell mass, including reticulocyte numbers and bone-marrow myeloid (granulocytes)-to-erythroid (red blood cells) (M : E) ratio (NC3Rs, 2016). Anaesthetising for blood sampling can also affect clinical pathology results (Bennett et al., 2009).

Nonstandardised blood-sampling methods and/or blood sampling by inadequately trained staff can result in artefacts such as haemolysis (the rupturing of red blood cells and the release of their contents into surrounding fluid), which can affect entire data sets, and introduce animal stress-related changes, which can result in variations in some variables (e.g. leukocyte counts), again affecting data interpretation (Gunn-Christie et al., 2012; Vap et al., 2012).

Some artefacts that can be introduced during the preanalytic phase (e.g. due to partial clotting of samples that were difficult to collect from very young or small animals, inadequate aspiration technique (leading to haemolysis), the use of inappropriate tubes) resulting in incorrect sampling ratios between anticoagulant and blood volume (Gunn-Christie et al., 2012). The most important thing is to correctly document any incident that could potentially cause skewed clinical pathology data later on.

It is recommended that study personnel involve the clinical pathologist in the review of the final study plan. This should include details such as the total allowable blood volume to be removed per animal by the end of the treatment phase, the types of tubes used for blood collection, the most scientifically meaningful profile of analytes for the purpose of the study, and acceptable storage and transportation of samples (ambient temperature versus refrigerator, time until separation of serum, etc.). In addition, standard procedures, such as randomisation of the blood-sampling order and fasting of the animals, should be discussed and defined before initiating the study (Tomlinson et al., 2013).

In general, whole blood is collected in plain serum tubes (serum is the colourless fluid produced after centrifugation of clotted blood) for use in creating biochemical profiles. In some environments, the preferred practice is to collect blood into tubes containing lithium heparin, and to determine the biochemical profile on heparin plasma (plasma is the colourless fluid produced after centrifugation of blood and anticoagulant). There are likely small differences between serum and heparin plasma (with the exception of fibrinogen in plasma), but the validation method and reference intervals should be established on the relevant matrix. Matrix includes all the properties of blood components in a particular species that can affect test performance, e.g. serum colour can affect photometric read-out. Usually, this is a nondefined effect, but it will result in biased results when comparing a standard with a particular species-specific specimen. For haematology profiles, the preferred anticoagulant is ethylenediaminetetraacetic acid (EDTA). For coagulation tests, it is citrate.

If the use of special kits designed for human use is planned (e.g. for biomarkers), the laboratory should be allowed sufficient time to properly validate such assays (Jensen and Kjelgaard-Hansen, 2006), and pretreatment blood samples should be scheduled to allow adequate interpretation of potential individual test article-related changes.

6.1.2 Analytic Phase: Data Generation

The analytic phase includes the actual generation of clinical pathology data (i.e. the measurement of all variables and analytes) (Tomlinson et al., 2013). This phase also includes proper laboratory management practices, including method validation and the regular (e.g. daily, monthly) quality control of all applied testing methods and instruments (Flatland et al., 2014; Camus et al., 2015). The personnel performing the analytic phase in preclinical studies should have adequate training on all the instruments they operate, as well as on the interpretation of instrument flags (error messages) and their follow-up (e.g. evaluation of blood smears in haematology, or trouble-shooting of coagulation tests via understanding of the coagulation cascade, including the intrinsic, the extrinsic and the common pathways) (Tomlinson et al., 2013).

Although some of the instruments developed for clinical pathology are advertised by their manufacturers as being very user-friendly, they still require adequate training of the personnel operating them.

6.1.3 Postanalytic Phase: Data Interpretation and Reporting

The postanalytic phase consists of the actual assessment and reporting of all measured variables (Tomlinson et al., 2013). The individual assessing clinical pathology data should be trained in the recognition of potential artefacts or laboratory errors, and should have a good understanding of the physiologic, pharmacologic and pathologic context of test item-related changes of toxicologic significance.

The storage of leftover samples is a very important component of postanalytic management; this includes freezer management and the keeping of detailed freezer logs. Leftover samples are important not only as back-up for confirmation of questionable results or the addition of new variables, but also for validation purposes (Flatland et al., 2014).

In summary, a high degree of standardisation, as reflected in carefully updated standard operating procedures (SOPs), helps minimise artefacts and variability, allowing for efficient and scientifically meaningful data interpretation. Clearly, the three phases of clinical pathology are very much interdependent, and they must be coordinated accordingly, ideally with the input of a certified clinical pathologist (Figure 6.1) (Tomlinson et al., 2013).

6.1.4 Good Laboratory Practice

Good Laboratory Practice (GLP) standards apply to the clinical pathology laboratory. SOPs should be established for all routine procedures. Importantly, a large amount of

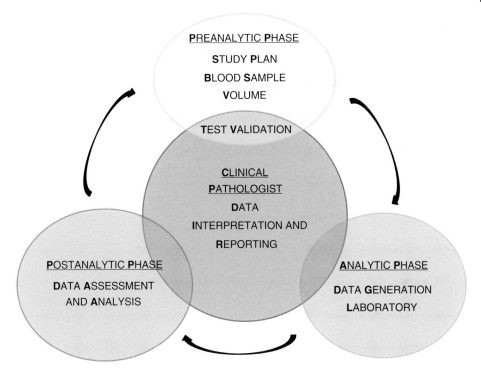

Figure 6.1 The three phases in clinical pathology.

raw data are generated in a clinical pathology facility. The handling of these data includes transfer, tabulation, statistical analysis and final archiving, all of which must follow validated procedures. This includes the validation of any laboratory informatics management systems (LIMS) and software. Details can be found in FDA (2015).

6.2   What is Measured in Clinical Pathology?

Blood is composed of a water base, which is termed 'plasma' or 'serum' depending on the separation of anticoagulated (plasma) or clotted (serum) blood from its cellular components (Figure 6.2). Serum is the main matrix for the measurement of the concentration of chemical variables (e.g. electrolytes, minerals and metabolites) and enzyme activities. Many of these analytes can provide information on the rapidly changing physiology or pathology of certain tissues and organs (e.g. heart, liver, kidney). In contrast, the cellular blood components (i.e. the red and white blood cells and platelets (thrombocytes)) are evaluated based on counts in whole anticoagulated blood and their morphologic appearance in a blood smear (Figure 6.2). Increased or decreased blood cell numbers can provide valuable information on potential test article-related effects such as blood loss, inhibition of haematopoiesis (blood cell formation in bone marrow) and inflammatory reactions. In addition, changes in morphology can point to disturbed maturation in the bone marrow or metabolic effects (e.g. phospholipidosis). Tables 6.1–6.4 provide a list of standard variables that should be evaluated in toxicologic

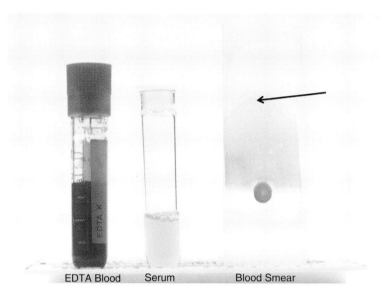

EDTA Blood   Serum   Blood Smear

Figure 6.2 The main materials for toxicologic clinical pathology testing are anticoagulated whole blood for haematology (e.g. in EDTA), serum collected after centrifugation of clotted whole blood for clinical chemistry and Wright–Giemsa-stained blood smears for microscopic evaluation and validation of quantitative and qualitative aspects of red and white blood cells and platelets. The arrow points to the feathered edge, where platelet clumps or atypical cells can preferentially be seen.

studies (Tomlinson et al., 2013). Notably, there are some species-specific differences. For instance, in rats and dogs, increased alanine aminotransferase (ALT) and aspartate aminotransferase (AST) activities are indicative of potential hepatotoxicity, whereas in pigs, increased sorbitol dehydrogenase (SDH) activity is more specific for hepatotoxicity (Tomlinson et al., 2013).

6.2.1 Interference by Haemolysis, Lipaemia and Icterus

Haemolysis (ruptured red blood cells), lipaemia (fat molecules in the blood) and icterus (yellow discolouration of tissues due to high blood bilirubin levels) are relatively common changes that can be either incidental or test article-related. They are visible as colour changes in plasma or serum after centrifugation of coagulated or anticoagulated blood (Figure 6.3). Such colour changes should always be carefully documented, as they can provide important information on either preanalytic issues or analytic concerns. For instance, haemolysis due to suboptimal blood-sampling practices requires a change in blood-sampling protocols, whereas haemolysis due to a test article-related effect (e.g. combined with a reduced red blood cell mass) requires in vitro haemolysis testing. Lipaemia may result from food consumption shortly before blood sampling; this requires fasting procedures prior to blood sampling. Alternatively, test article-related lipaemia can be indicative of metabolic changes. Finally, icterus is indicative of increased bilirubin concentrations (e.g. as a consequence of marked haemolysis or potential hepatobiliary dysfunction). Haemolysis, lipaemia and icterus can all interfere with some analytic tests. Each laboratory should establish its own list of potential interferences based on the assays it is using.

Table 6.1 Recommended haematology variables for routine toxicology studies

Variable	Unit
Red blood cell mass	
Haematocrit	%
Haemoglobin concentration	g/l
Red blood cell count	10^{12}/l
Indices	
Mean cell volume (MCV)	fl
Mean cell haemoglobin (MCH)	pg
Mean cell haemoglobin concentration (MCHC)	g/l
Reticulocyte count	
Red cell distribution width (RDW)	%
Total white blood cell count	10^{9}/l
Differential white blood cell count (absolute counts only)	
Neutrophils	10^{9}/l
Lymphocytes	10^{9}/l
Monocytes	10^{9}/l
Eosinophils	10^{9}/l
Basophils	10^{9}/l
Large unclassified cells (LUCs)	10^{9}/l

Table 6.2 Recommended coagulation variables for routine toxicology studies

Variable	Unit
Prothrombin time (PT)	seconds
Activated partial thromboplastin time (APTT)	seconds
Platelet count	10^{9}/l
Fibrinogen	mg/l

6.3 Haematology

Haematology primarily evaluates the cellular components of blood for their qualitative and quantitative properties. Qualitative or morphologic aspects include the recognition of maturation status (e.g. polychromasia: different-coloured red blood cells; left shift: release of immature neutrophils), signs of toxic change or activation (e.g. in neutrophils, lymphoblasts, large unclassified cells (LUCs)) and the presence of other morphologic changes (e.g. intracytoplasmic vacuoles in phospholipidosis) in the different cell populations. There are three main cell populations, which also represent the developmental

Table 6.3 Recommended serum chemistry analytes for routine toxicology studies

Analyte	Unit
Enzymes	
Aspartate aminotransferase (AST)	U/l
Alanine aminotransferase (ALT)	U/l
Sorbitol dehydrogenase (SDH) (minipigs)	U/l
Glutamate dehydrogenase (GLDH) (minipigs)	U/l
Alkaline phosphatase (ALP)	U/l
γ-glutamyltransferase (GGT) (rodents)	U/l
Creatine kinase (CK)	
Metabolites	
Total bilirubin	μmol/l
Glucose	mmol/l
Urea	mmol/l
Creatinine	μmol/l
Triglycerides	mmol/l
Total cholesterol	mmol/l
Total protein	g/l
Albumin	g/l
Albumin : globulin ratio	g/g
Electrolytes	
Sodium	mmol/l
Potassium	mmol/l
Chloride	mmol/l
Total calcium	mmol/l
Phosphorus	mmol/l

lineages originating in the bone marrow: red blood cells or erythrocytes, white blood cells or leukocytes and platelets or thrombocytes.

For more detailed information concerning any morphologic or quantitative aspect of blood cells, it is recommended to consult a standard veterinary haematology reference, such as Weiss and Wardrop (2010).

6.3.1 Manual and Automated Techniques in Haematology

Over the last decade, automated haematology analysers equipped with species-specific software for most of the species used in preclinical studies have become standard in most laboratories. Although these instruments provide reliable and consistent results, errors due to preanalytic incidents or biologic aberrations require the operating personnel to be familiar with traditional manual haematology practices; they must be able to confirm atypical results indicated by instrument flags or error warnings by assessing red and

Table 6.4 Recommended urinalysis variables for routine toxicology studies

Variable	Unit
Volume	ml
Colour	Macroscopic description
Clarity	Macroscopic description
Specific gravity	g/g
pH	
Glucose	−, +, ++, +++
Protein	−, +, ++, +++
Blood	−, +, ++, +++
Ketones	−, +, ++, +++
Bilirubin	−, +, ++, +++

white blood cell counts by alternative means, and to evaluate blood smears for morphologic changes, including common artefacts such as platelet clumps.

6.3.2 Haematocrit and Red Blood Cell Mass

The haematocrit, or packed cell volume (PCV), is a quantification of the cellular components relative to the liquid phase of whole or anticoagulated blood (Figure 6.3). The proper term for instrument-generated data is 'haematocrit', which is extrapolated from flow cytometrically determined, mean cell volume (MCV) and red blood cell counts. Manually, the PCV is determined in a microhaematocrit tube after centrifugation to

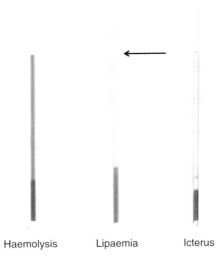

Haemolysis   Lipaemia   Icterus

Figure 6.3 Microhaematocrit tubes with haemolysis, lipaemia (small fat layer after centrifugation; arrow) and marked icterus in the plasma layer. Such discolourations can be accidental or treatment-related and can interfere with analytical measurements. *Source:* Courtesy Dr Ilse Schwendenwein, University of Veterinary Medicine Vienna.

separate the cells from plasma (Figure 6.3). As a more functional unit, the red blood cell mass may include all variables providing information on the total oxygen-carrying capacity of peripheral blood, including haematocrit/PCV, red blood cell count (including reticulocyte count), haemoglobin concentration and the indices (see Section 6.3.3.1). A reduction in red blood cell mass is usually accompanied by a decrease of haematocrit/PCV, red blood cell count and haemoglobin concentration, while reticulocyte count and indices can vary according to the origin and pathophysiology of the reduction.

6.3.3 Blood Cells

6.3.3.1 Red Blood Cells

Mammalian red blood cells are biconcave, discoid anuclear cells that transport haemoglobin – the protein responsible for binding oxygen in the lung and transporting it to the tissues. Oxygen binding is mediated by iron, a potentially highly oxidative and toxic element contained inside the porphyrine ring, which is held in position by four globin molecules. A basic understanding of haemoglobin synthesis and function can be of relevance when evaluating test article-related effects on red blood cell mass in many different settings. For more details, see Weiss and Wardrop (2010).

In mature animals, red blood cells originate in erythropoietic islands in bone marrow, which are present in the marrow cavity of trabecular bone. The earliest precursor cell, the erythroblast, is a large basophilic cell with a large amount of dark-blue cytoplasm and a large nucleus with a nucleolus. Transferrin receptors are expressed on maturing erythroid precursor cells, allowing the transfer of iron and the synthesis of haemoglobin, which results in an increasing orange-red colouration of the cytoplasm in routine stains of the subsequent stages of red blood cells, which also decrease in size with advanced maturation and haemoglobination. At the same time, the nucleus condenses continuously until only a small, compact, round, dark-blue or black apoptotic body remains inside the metarubricyte. Shortly before the reticulocyte or polychromatic immature red blood cell is released into the peripheral blood circulation, the apoptotic nucleus is expelled and phagocytised by macrophages. Immature red blood cells may still contain some remnants of RNA, giving them a slightly bluish tinge (polychromasia) in routine stains (Figure 6.4). The term 'reticulocyte' comes from the appearance of such young red blood cells after staining with new methylene blue, which demonstrates a network of ribosomes, appearing as a reticular or network pattern. Examination of new methylene blue-stained blood smears is the traditional method for assessing reticulocytes (Figure 6.4) (Weiss and Wardrop, 2010).

In cases of acute need, such as following acute haemorrhage or intense blood-sampling regimens, the release of immature red blood cells can be accelerated such that increasing numbers of reticulocytes or polychromatic cells, or even nucleated red blood cells (nRBCs) or metarubricytes, will be visible in blood smears. In addition, the mean cellular volume (MCV) will be increased, and red cell distribution width (RDW) will be enlarged or even biphasic, because the immature precursors released in response to the increased demand will be larger than the mature erythrocyte (Figure 6.4).

Modern haematology analysers report reticulocytes based on the amount of stainable cytoplasmic RNA in absolute and relative numbers (as a percentage of total red blood cells). In addition, nRBCs are reported. Typically, regenerative haematology profiles are accompanied by an increased MCV, manifesting as anisocytosis (variation in cell size) in the blood smear (as immature red blood cells are slightly larger, and thus there are erythrocytes of different sizes present), which in a toxicologic clinical pathology context is usually considered a normal physiologic response to blood loss (Figure 6.4).

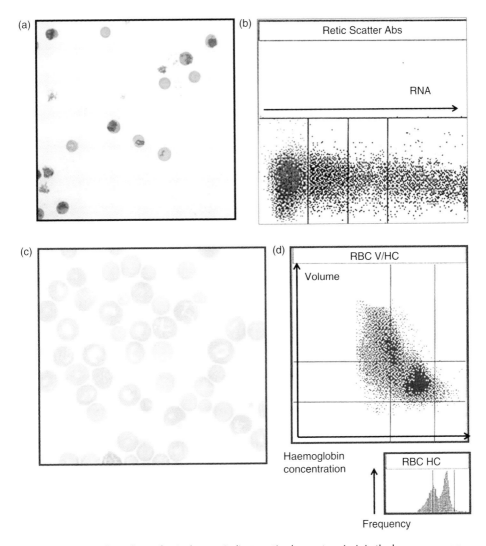

Figure 6.4 Increased numbers of reticulocytes indicate active haematopoiesis in the bone marrow, a physiologic response to blood loss due to moderate or excessive blood sampling, haemorrhage or haemolysis. (a) Traditionally, reticulocytes are visualised by vital stains, such as new methylene blue, which make the reticular pattern of remnant ribosomes and RNA visible. (b) In the haematology analyser, reticulocytes are identified by their remnant RNA (in the absence of a nucleus) (x-axis), with most immature reticulocytes on the far-right side. (c) In a Wright–Giemsa-stained blood smear, younger red blood cells are characterised by a larger diameter (anisocytosis) and the lack of a central pallor, as well as a slightly bluish tinge, termed 'polychromasia'. (d) In the analyser, these cells typically have a higher volume (y-axis) and a lower haemoglobin concentration (x-axis). In a very pronounced response, there can actually be two populations of red blood cells, with normal and lower haemoglobin concentrations (x-axis). *Source*: Courtesy Dr Ilse Schwendenwein, University of Veterinary Medicine Vienna.

The size and haemoglobin content of individual red blood cells is assessed by the indices: the MCV, the mean cell haemoglobin (MCH) and the mean cell haemoglobin concentration (MCHC). More recently, a new variable, the RDW, is also being reported by modern haematology analysers. This represents a histogram of the frequency distribution of the differently sized, red blood cells. In regenerative red-blood-cell profiles, the MCV and RDW will typically be increased, due to the appearance of larger (macrocytic) immature red blood cells (Figure 6.4). Depending on the availability of iron for haemoglobin synthesis, MCH and MCHC can be slightly lower, especially in cases of chronic ongoing inflammatory processes (e.g. anaemia of chronic disease).

If the location of blood loss is not evident from either in-life observations or excessive blood sampling for analytic purposes, coagulation profiles should be checked for a potential explanation (e.g. coagulation disorder causing bleeding and blood loss).

6.3.3.2 White Blood Cells

White blood cells are responsible for antigen-specific (lymphocytes) and -nonspecific (granulocytes, monocytes) defences against infectious pathogens. While granulocytes contain cytoplasmic granules with cytotoxic and vasoactive compounds (peroxidase, histamine and many others), they are also capable of phagocytosis of bacteria – which is also the main function of activated monocytes or macrophages. The numbers, migration and actual physiology of white blood cells in blood and tissues are regulated by complex networks of messenger molecules (i.e. cytokines and immunoregulatory networks), which can be modulated by treatments (particularly biologicals), leading to test article-related changes in leukocyte numbers in the absence of infectious pathogens. Importantly, test article-related effects concerning the absolute and differential numbers of white blood cells may not be apparent in the histopathology analysis, and thus clinical pathology data may provide the only indication of a treatment effect.

There are three types of granulocytes: the neutrophil, eosinophil and basophil. All granulocytes are characterised by lobulated or convoluted to segmented nuclei with a moderate amount of cytoplasm, which contains different types of granules that either do not stain (neutrophil) or take on a pink (eosinophil) or purple to blue (basophil) colour after staining with standard Romanowsky stains – hence the descriptive terminology (Figure 6.5). The predominant granulocyte is the neutrophil, while the eosinophil can be relevant in particular immunologic settings, such as allergic reactions and parasitic infections (particularly in non-human primates). The number of eosinophils is typically lower in the presence of an adrenocortical stress reaction (see later), while the neutrophil count will be higher (Bennett et al., 2009; Everds et al., 2013). The basophil is in general of no relevance in toxicologic clinical pathology, but may parallel changes in eosinophil numbers. Like red blood cells, the granulocytes originate in the bone marrow, where they undergo a maturation process from the early myeloblast with an agranular cytoplasm and a large nucleus with a nucleolus to the mature segmented granulocyte with granular cytoplasm and a segmented condensed nucleus. In times of increased need, such as under the influence of certain cytokines, severe inflammation or other mediators, a left shift – characterised by the premature release of band neutrophils – can be seen (Figure 6.5). By definition, the nucleus of the band neutrophil displays parallel outlines and no constrictions, as seen in the mature neutrophil. Changes in neutrophil counts can happen relatively swiftly due to either release from the wall of blood vessels during stress, extravasation to tissues due to chemotaxis or rapid release from bone marrow.

Lymphocytes are produced in the bone marrow, but also originate in the thymus in growing animals and in the lymph nodes and other secondary lymphoid tissue in the body (e.g. mucosal-associated lymphoid tissue (MALT), gut-associated lymphoid tissue (GALT)). Lymphocytes generally appear as round cells with a small rim of basophilic

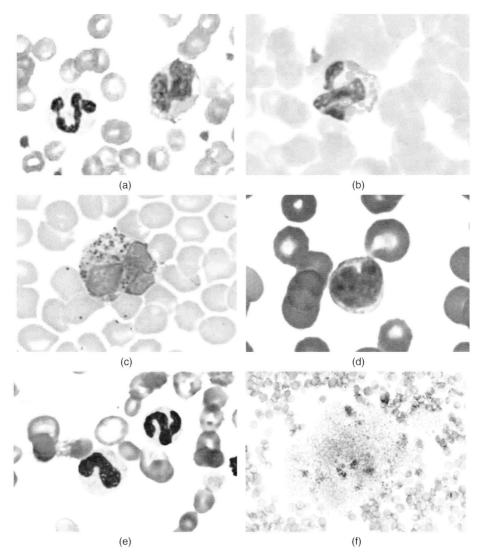

(a)

(b)

(c)

(d)

(e)

(f)

Figure 6.5 (a–e) dog; (f) cat. Leukocytes and platelets in canine blood smear. Wright–Giemsa, ×100 objective (a–e). (a) Segmented neutrophil with pale cytoplasm and an irregular convoluted and knobby condensed nucleus (left); monocyte with slightly basophilic/bluish cytoplasm and an irregular shaped nucleus with a reticulated chromatin pattern (right). (b) Eosinophil with large eosinophilic cytoplasmic granules and an irregular ribbon-shaped nucleus, (c) Basophil with numerous purple cytoplasmic granules and an irregular lobulated nucleus. (d) Lymphocyte with a small rim of deeply basophilic/blue cytoplasm and a dense bean-shaped nucleus. (e) Segmented (right) and band (left) neutrophil with pale cytoplasm and the typical band-shaped nucleus. (f) Platelet aggregates ×20 objective. *Source*: Courtesy Dr Ilse Schwendenwein, University of Veterinary Medicine Vienna.

cytoplasm and usually round nuclei without visible nucleoli. They are slightly smaller than granulocytes, but larger than red blood cells (Figure 6.5). After stimulation of the immune system, either by cytokines or by immune-modulating compounds, activated lymphocytes can transform into lymphoblasts, which are characterised by increased cellular and nuclear size, and sometimes by visible nucleoli. Lymphocyte numbers can be decreased under the influence of an adrenocortical stress reaction, which is usually accompanied by a decrease in the size and weight of immunologic organs (Everds et al., 2013).

Lymphocytes can be further classified based on immunophenotyping, where superficial or cytoplasmic antigens or cluster-of-differentiation (CD) markers are labelled by specific antibodies. In addition to B- and T-lymphocytes, many other phenotypes have been characterised, including helper and suppressor, cytotoxic and regulatory cell types. The most commonly used antibodies label CD4 (T-helper) or CD8 (T-suppressor and T-cytotoxic) surface antigens. Special immunology texts, such as Abbas and Lichtman (2014) and Rütgen et al. (2015), should be consulted for further information on immunophenotyping.

Monocytes (or macrophages, when activated) represent a smaller but equally important type of leukocyte. They originate in the bone marrow and arise from the myeloid cell lineage. They are relatively large cells, with pale, nongranulated cytoplasm and a pleomorphic nucleus that can be round or band-shaped (Figure 6.5), and are specialised for phagocytosis and storage of iron. In addition, they can secrete cytokines. Monocyte counts often increase together with neutrophil or lymphocyte counts, indicating a possible inflammatory response.

6.3.3.3 Platelets

Mammalian platelets, or thrombocytes, are small, anuclear, granulated cells that play a major role in coagulation and in some inflammatory processes. They are generally smaller than other blood cells, and tend to aggregate into small clumps in some species (particularly the rat), even in the presence of anticoagulants. Such clumps can be visible on the feathered edge of a standard blood smear (Figures 6.2 and 6.5f), the evaluation of which should be standard practice in any situation in which the haematology analyser displays error flags or abnormal results for platelet counts. Activated platelets release a number of molecules that play a role in coagulation and the maintenance of integrity of vessel walls (Hoffman and Monroe, 2001).

Platelets originate from megakaryocytes in bone marrow. These are large, multinucleated cells with a large amount of sometimes finely granular cytoplasm (book cover figure).

Lower numbers of platelets can be caused by either increased loss (e.g. by increased consumption or blood loss) or reduced production in bone marrow. An important artefact is platelet clumping, which causes falsely reduced platelet counts. A microscopic evaluation of the number and morphology of bone-marrow megakaryocytes either in a smear or on a histologic section of bone marrow will allow the distinction between a production deficit or increased peripheral need due to destruction and loss of platelets.

6.3.4 The Standard Haematology Profile

The standard haematology profile, as it is produced by most modern analysers, includes variables that provide information on the number and maturity of red and white blood cells (Table 6.1) (Tomlinson et al., 2013). The number and morphology of platelets can be reported as part of either the haematology profile or the coagulation profile (Table 6.2). In addition, it is recommended that blood smears be collected in all studies (Tomlinson et al., 2013). Archived smears can be analysed at a later time point, in case of need. Proper interpretation of abnormal findings requires solid training of personnel in the normal red and white blood cell morphology in all species used in preclinical studies.

6.3.5 Bone Marrow

The bone marrow is the site of haemato- and lymphopoiesis, and, as such, is the origin of most normal and abnormal blood cells. Detailed recommendations on the performance of bone-marrow examination have been published by Reagan et al. (2011). Cytologic myelogram, or bone-marrow smear examination, by an expert clinical pathologist is a resource-intensive exercise requiring substantial amounts of time and commitment. Therefore, although collection of bone-marrow smears at necropsy is standard practice in most preclinical studies, the actual myelogram should only be performed according to relatively stringent indications (Reagan et al., 2011).

Bone-marrow examination should be performed in studies with clear treatment-related changes in haematology variables if the following apply:

- Test article-related changes in haematology variables cannot be explained by frequently seen mechanisms, such as loss of red blood cell mass due to chronic inflammation or disease, or excessive blood sampling or haemorrhage (see Sections 6.15.1 and 6.15.3).
- There are no changes in body weight, body weight gain or food consumption that could explain test article-related effects in haematology variables.
- There are quantitative and/or morphologic changes in all three cell lines (erythroid, myeloid and thromboid).
- The test item is known to specifically affect haematopoiesis in the bone marrow.

6.4   Coagulation

Coagulation, or blood clotting, is a normal physiologic component of haemostasis, a process initiated after injury to a blood vessel. Haemostasis consists of the sequential activation of a series of enzymes (originally called 'the clotting cascade'), including the intrinsic and the extrinsic systems, leading to the formation of fibrin strands, which ultimately build a thrombus or clot to seal off the vessel wall (Figure 6.6a) (Kurata and Horii, 2004). Recently, it has been emphasised that this is not just a proteolytic process, but also a cellular one, as fibroblasts, endothelial cells and activated platelets are responsible for the actual release of components (such as tissue thromboplastin, phospholipid and calcium) that initiate the coagulation cascade (Figure 6.6b,c) (Hoffman and Monroe, 2001). For analytic purposes, blood collected for coagulation testing is anticoagulated with a strictly defined volume of sodium citrate (nine parts sodium citrate and one part blood), which prevents activation of the coagulation cascade in the tube by binding the available calcium. Analysers require either plasma removed after centrifugation of citrated blood or citrated whole blood. After resupplying calcium and a coagulation activator such as recombinant human thromboplastin or phospholipid, actual clotting is usually determined in seconds based on the formation of fibrin. Increased real coagulation

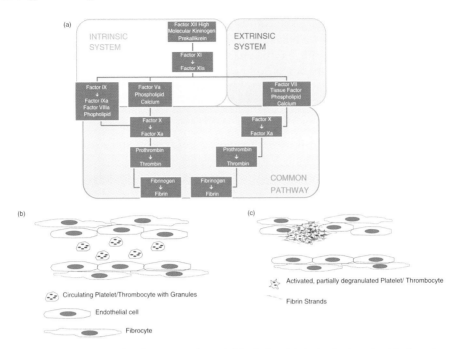

Figure 6.6 Coagulation in the classic cascade model, including (a) the intrinsic and extrinsic systems and the common pathway, (b) the main cellular contributors to clot formation, endothelial cells and platelets and (c) thrombotic clot formation with the participation of activated platelets and crosslinked fibrin strands in haemostasis following disruption of a vessel wall. The cell-based model for coagulation can be seen in Hoffman and Monroe (2001).

times therefore indicate a change in coagulation factors – usually a decrease – resulting in a prolonged or deficient process of coagulation. As many coagulation factors are synthesised in the liver, deficient coagulation should always raise suspicion of test article-related hepatic dysfunction.

In many laboratories, fibrinogen is determined as a particular coagulation variable by the method of Clauss (derived from prothrombin time (PT)) (Ameri et al., 2011). However, a significant increase in fibrinogen is a marker for an acute-phase reaction (i.e. an actual inflammatory response). In such cases, it is recommended that other species-specific, acute-phase markers (e.g. C-reactive protein (CRP) (dogs), α1 acid glycoprotein (rats), serum Amyloid A (non-human primates)) be measured to confirm an inflammatory response.

Another technique for the assessment of haemostasis is thromboelastography, which allows the combined measurement of coagulation factors and platelet function. Although thromboelastography is a standard test in intensive-care and clinical settings, it is not considered standard in toxicologic clinical pathology, due to the lack of a high-throughput format and the need for very high and reproducible sample quality (Kurata and Horii, 2004).

6.4.1 Standard Coagulation Profile

The standard coagulation profile includes the PT, the activated partial thromboplastin time (APTT) and the concentration of fibrinogen (Table 6.2). In addition, platelets are often reported as a coagulation variable. The PT and APTT are each dependent on the addition (during testing) of different activators, including calcium and factors usually designed for the initiation of human blood coagulation (e.g. recombinant tissue factor). Therefore, careful validation and quality control of such reagents is required for each species tested, as well as each analyser, and for specific reagents and their batches. This is particularly relevant when PT and APTT results are expected to fall outside of the normal human range, as may be the case for canine specimens or samples from animals with test article-related changes in coagulation times (e.g. drugs against thrombosis). In such circumstances, instrument settings need to be adjusted to the special needs of the species and the study. For details on the coagulation cascade and all the different factors involved in it, see Kurata and Horii (2004).

6.4.2 Prothrombin Time

The PT is an assessment of the extrinsic portion of the clotting cascade, particularly factors VII and X. It typically represents the coagulation reaction elicited by tissue trauma (Figure 6.6a). The PT is always shorter than APTT, and the canine PT is much shorter than human PT (requiring proper setting of automated equipment designed for human use).

6.4.3 Activated Partial Thromboplastin Time

The APTT is an assessment of the intrinsic portion of the clotting cascade, particularly factors XII, XI, IX and X (Figure 6.6a). It represents the coagulation reaction elicited by contact activators and is always longer than PT.

6.4.4 Fibrinogen

Fibrinogen can be extrapolated from PT. Alternatively, immunologic and physical analytic methods are available (Ameri et al., 2011).

6.5 Clinical Chemistry

Clinical chemistry investigates the analytes present in serum after centrifugation of coagulated blood (Figure 6.2); alternatively, analytes can also be investigated in plasma from blood collected in lithium heparin. The analytes of interest include a number of enzyme activities, electrolytes, minerals and concentrations of components of protein, carbohydrate and fat metabolism, all of which can provide valuable information on overall body condition and organ function and integrity. For more detailed information, see Kaneko et al. (2008).

6.5.1 Metabolites

6.5.1.1 Carbohydrate Metabolism

Glucose is the main energy source of common laboratory species, either directly absorbed from the intestine or synthesised in the liver. The concentration of blood glucose is regulated by a complex endocrine network. Glucose will continue to be consumed by live cells after blood collection; therefore, blood-glucose concentration tends to be lower in samples that have been stored for several hours. If test item-related changes in blood glucose are anticipated, either point-of-care instrumentation should be considered (after careful validation of the analytic range in question) or stringent blood separation and storage protocols should be followed.

Marked increases or decreases in blood sugar (hyper- or hypoglycaemia) are life-threatening conditions that can be seen in dogs and monkeys with test item-related endocrine disorders, such as subnormal function of the cortex of the adrenal gland (hypoadrenocorticism) and diabetes mellitus, respectively. The abnormal blood-sugar concentrations can be accompanied by an increase in ketone bodies in the urine and a drop in blood pH (metabolic acidosis). In addition, increased excretion of glucose in urine (diabetes mellitus) can be accompanied by dehydration and a tendency for urinary tract infections. Moderate hyperglycaemia is typically seen in rats with a stress reaction (Everds et al., 2013).

6.5.1.2 Protein Metabolism

Proteins represent the building blocks of all body tissues. Protein metabolism is usually assessed by measuring the total concentration of protein and of its main component albumin, while globulins represent the calculated difference after deduction of the concentration of albumin from total protein. The albumin-to-globulin (A : G) ratio can provide information on an imbalance between the main measured protein components. Advanced analysis of different globulin fractions is carried out by serum protein electrophoresis.

Decreased protein concentrations involving both albumin and globulins usually suggest compromised uptake (e.g. lower food consumption) or increased gastrointestinal loss (e.g. diarrhoea). Decreased albumin levels should raise concern about an acute-phase reaction, hepatic dysfunction or renal loss. Increased globulin concentrations can be indicative of an acute-phase response or an immunologic reaction (due to increased γ-globulins or immunoglobulins). For more detail on immunoglobulin metabolism, see Abbas and Lichtman (2014).

Protein degradation ultimately results in urea, which is synthesised in the liver and cleared by the kidneys (see Section 6.5.4.2). A test item-related decrease of urea suggests liver dysfunction, while increased urea can be caused by dehydration and haemo-concentration, in which case it is accompanied by increased erythroid mass variables and albumin concentration, or by renal dysfunction. In the latter case, an increase in creatinine may also be observed, and the specific gravity in urine may be lower than normal (see Section 6.6).

6.5.1.3 Lipid Metabolism

Lipids represent an important storable source of energy and are a component of membrane structures in the body. The main variables measured in preclinical studies for assessment of lipid metabolism are concentrations of cholesterol and triglycerides. Subgroups of cholesterol such as high- and low-density lipoproteins (HDL, LDL) are sometimes determined in studies testing compounds that affect lipid metabolism. As for other nonstandard analytes, assays determining HDL (and calculated LDL) should be carefully validated for the respective species before use of such data in preclinical studies.

Lipids can be absorbed directly from the intestinal tract, and metabolism mostly takes place in the liver under the influence of the endocrine network. Changes in lipid concentrations may therefore indicate decreased food intake or an alteration of hepatic function. However, initially, careful definition of the fasting or nonfasting status of animals prior to blood sampling is relevant for lipid determination in peripheral blood. For instance, the presence of postprandial (i.e. after feeding) lipaemia can severely affect the measurement of a number of analytes, due to interference with the laboratory methods.

6.5.1.4 Metabolism of Haemoglobin

While the rate of haemoglobin synthesis is an important marker of erythropoiesis, the degradation products of haemoglobin – bilirubin and its conjugated and unconjugated forms – are markers of hepatobiliary function (because they are excreted by the liver), and the concentration of different forms of iron can be relevant under circumstances of chronic inflammation or a protracted acute-phase response (Cray et al., 2009).

An increase in total bilirubin concentration suggests hepatobiliary dysfunction or a haemolytic process. A separate determination of direct (conjugated in liver, increased in biliary stasis) or indirect (unconjugated, increased in hepatic dysfunction) bilirubin may be warranted, although the increase in conjugated bilirubin, which is water-soluble, can often be confirmed by increased bilirubin in urine.

6.5.2 Enzymes

There are a number of enzymes that are assessed based on measured activity. In general, it is assumed that significantly increased enzyme activities are consistent with increased cellular release, usually due to a loss of cell integrity. However, increased enzyme activity can also be caused by induction of enzyme gene expression and transcription (e.g. alkaline phosphatase (ALP) isoenzyme induction in dogs with adrenocortical hormone or test item-related induction of ALT isoenzymes) (Everds et al., 2013). In addition, ALP activity is higher in growing animals, due to the release of the bone isoenzyme by proliferating osteoblasts. Another important aspect is the clearance of enzymes from peripheral blood. The average half-life of different isoenzymes has not been determined in all species, but with a half-life of 60 hours, acute massive liver necrosis after a single treatment may be accompanied by normal ALT activities by day 7 of a study (Ennulat et al., 2010).

The most commonly assessed enzyme activities include markers for hepatic integrity (ALT and AST) and hepatobiliary function (ALP); in some species, such as pigs (SDH and glutamate dehydrogenase (GLDH)) and rodents (γ-glutamyltransferase (GGT)), alternative enzyme activities can be measured (Tomlinson et al., 2013). Finally, creatine kinase (CK) is a marker for striated muscle integrity, including both skeletal and cardiac muscle.

'Organ dysfunction' and 'organ integrity' are not equivalent terms and are investigated with different enzyme and metabolite profiles. In addition, a final assessment generally requires confirmation by histopathologic examination (e.g. confirmation of hepatic necrosis).

6.5.3 Electrolytes and Minerals

Electrolytes, such as sodium, potassium and chloride, are sensitive indicators of overall hydration balance and renal function. In contrast, minerals, such as calcium, phosphorus and magnesium, are only present in a small proportion in the blood when compared to total body concentration.

6.5.3.1 Potassium

Potassium is present in high concentrations inside cells; therefore, any type of massive cell destruction (haemolysis, muscle necrosis) can result in transiently increased potassium concentrations in peripheral blood. In such cases, an accompanying hyperbilirubinaemia or increased CK would be expected. However, with normal renal function, the concentration of potassium in the serum should be rapidly restored to normal.

Changes in potassium concentration suggest renal dysfunction, and changes in the sodium-to-potassium ratio can be seen in adrenocortical hypofunction.

6.5.3.2 Sodium and Chloride

Sodium and chloride concentrations are usually lower in animals with protracted vomiting and diarrhoea and/or with profuse salivation. Therefore, consideration of clinical observations is always warranted in such cases.

6.5.3.3 Calcium and Phosphorus

Calcium and phosphorus are important components of bone metabolism, and may vary somewhat with the age of the animal. Since the majority of calcium is bound to albumin for transport in serum, most changes in calcium concentration are related to changes in albumin concentration. Phosphorus is lower in cases of reduced food intake. Calcium and phosphorus can change in cases of renal dysfunction, but not in isolation – generally, changes occur in association with increased urea and creatinine concentrations.

6.5.4 Standard Chemistry Profiles

The recommended standard chemistry profile is very similar in most species used in preclinical studies, and differs slightly from profiles used in clinical diagnostics. In general, it is recommended that one avoid variables that have not been well defined in terms of pathophysiologic significance, as changes in such analytes can present a challenge in data interpretation. One way to facilitate data interpretation is to group analytes in organ profiles.

6.5.4.1 Assessment of Liver Function

A typical liver profile should include liver enzyme activities such as ALT, AST, ALP and SDH, as well as GLDH in pigs and GGT in rodents, to provide information on hepatobiliary integrity. In addition, concentrations of metabolites such as total protein, albumin, urea, bilirubin, triglycerides and cholesterol, complemented by the coagulation profiles, can yield information on hepatobiliary function. Liver assessment usually needs to include histopathology findings in order to provide a conclusion on potential test article-related adverse effects.

6.5.4.2 Assessment of Kidney Function

A typical kidney profile should include blood urea and creatinine concentrations, providing preliminary information on clearance function, complemented by urinalysis variables (including specific gravity, indicating the kidney's concentrating ability) and other physical variables, such as the presence of protein, blood and leukocytes in the urine. The presence of these elements in urine may indicate test-related loss of kidney integrity (protein or blood loss) or inflammation (leukocytes). Protracted renal excretion of protein will result in lower serum albumin concentrations, and massive damage to the kidney could potentially reduce erythropoiesis, as erythropoietin is produced in the kidney.

While a combination of increased urea and creatinine serum concentrations with low urine specific gravity is diagnostic for renal dysfunction in clinical medicine, such patterns are rarely encountered in preclinical studies. For this reason, a major campaign for the validation of renal biomarkers was launched a few years ago.

6.5.4.3 Assessment of Gastrointestinal Function

Test item-related changes in the gastrointestinal tract are usually characterised by clinical signs such as vomiting, diarrhoea, salivation and/or decreased food consumption. Nevertheless, lower concentrations of albumin and globulins, triglycerides and cholesterol or sodium and chloride can accompany such changes. Associated dehydration is usually characterised by increased blood urea concentration with or without increased red-blood-cell mass variables. Changes in these analytes should always initiate a request for in-life observations.

6.6 Urinalysis

Urinalysis is part of the standard clinical pathology database in nonclinical studies (Tomlinson et al., 2013). Urine variables include physical variables such as volume, colour, pH and specific gravity. They provide preliminary information on renal function and the hydration status of an animal, as well as a semiquantitative evaluation of the physiologic and pathologic components of the urine, such as the presence of blood, red and white blood cells, protein, glucose, bilirubin and ketone bodies (Table 6.4). In addition to the physicochemical analysis, a microscopic evaluation of urinary sediment for confirmation of cellular or crystalloid particles can be appropriate in some studies (Tomlinson et al., 2013).

6.7 Acute-Phase Proteins

Acute-phase proteins include a group of proteins expressed and secreted mostly in the liver after an inflammatory stimulus (Cray et al., 2009). An acute-phase response can be initiated by many different aetiologies, including trauma, infection and a number of test items, particularly biologicals. Importantly, an acute-phase reaction may not be evident from any other noticeable changes in study variables (including the absence of histologic changes), except in assays measuring acute-phase proteins. Often, increased concentrations of fibrinogen may be the only hint (together with lower albumin concentrations) in a standard clinical pathology profile of the presence of an acute phase-response.

Serum concentrations of acute-phase proteins are usually measured with immunologic assays, which may or may not be species-specific. In most cases, especially if they are not species-specific, careful assay validation is recommended. Also, as changes in acute-phase proteins may be very dynamic over time, it is recommended to include

Table 6.5 Acute-phase proteins that show measurable changes in different laboratory animal species (Honjo et al., 2010; Heegaard et al., 2013; Christensen et al., 2014).

Species	C-reactive protein (CRP)	Serum amyloid A	A1 acid glycoprotein	A2-macroglobulin	Haptoglobin	Fibrinogen	Albumin[a]
Dog	×	×			×	×	×↓
Rat			×	×	×	×	×↓
Mouse		×	×		×		
Monkey	×	×		×		×	×↓
Pig	×	×				×	×↓

[a]Albumin is a negative acute-phase protein: its serum concentration is lower in an acute-phase reaction.

pretreatment samples in the evaluation and to measure those acute-phase proteins that have been found to undergo significant changes in a particular species (see Table 6.5).

6.8 The Biomarker Concept

Biomarkers have become popular recently, largely because of the expectation that they will allow the identification of organ-specific, test article-related adverse effects earlier in a study and at lower treatment doses than traditional clinical pathology variables. However, the increased sensitivity of such biomarkers often comes with a decrease in specificity. Unless very careful and well-designed studies (which include phenotypic anchoring of particular markers) are performed, caution is advised regarding the incorporation of insufficiently characterised and unvalidated biomarkers (Burkhardt et al., 2011). While some biomarkers may have value in studies involving preclinical research and development, their validity for incorporation into regulatory studies must be determined based on the most current scientific literature. The following minimal requirements should be fulfilled before a particular biomarker assay is included in the clinical pathology database:

- There should be a minimal database on analytic validation, including precision, accuracy, linearity under dilution and matrix effect-testing for each species investigated.
- Changes in biomarkers should be strictly related to pretreatment reference data compiled from the animals treated in the study.
- A control group should always be included as an additional normal reference group.
- Ideally, the sensitivity and specificity of a given biomarker should be assessed, and potential thresholds for toxicologic relevance should be determined based on receiver operator curves (ROCs) (Burkhardt et al., 2011).

Biomarkers can be valuable when taking a case-by-case approach, particularly in early research-and-development studies. In studies following GLP guidance, all generated data must be reported; thus, the biomarkers measured should be more or less established as diagnostic indicators, otherwise interpretation of marginal test item related changes can be very challenging.

6.9 Reference Intervals

Reference intervals are the normal range of values for a particular enzyme activity, cell count or analyte concentration, and by definition should be determined from a group of healthy or normal reference individuals (Friedrichs et al., 2012). They can be valuable, especially when assessing clinical pathology data from individual animals, during pre-treatment or in situations where test article-related changes are equivocal. They are also mandatory in target animal safety-type studies. However, reference intervals are only valid if they are regularly updated and compiled using specimens taken from healthy animals, either from the control group or from all animals during the pretreatment phase. The reference animal population should be of the same species, strain, origin, sex and age group as the treated animals (Geffre et al., 2009). A number of articles have been published recently addressing best practices in reference interval determination, including for small (<120 individuals) groups of reference animals (Geffre et al., 2011; Braun et al., 2013). Such reference intervals can be established using recommended statistical analysis methods. If only small numbers of animals are available as a normal or healthy reference population, then reference limits can be used instead of actual intervals (Geffre et al., 2011; Braun et al., 2013).

It is recommended that study personnel consult key publications before establishing reference intervals. Using reference intervals published in books or generated at partner laboratories might be acceptable if they have been validated with a small sample population in that particular institution, but otherwise it is considered poor practice.

6.10 Instrumentation, Validation and Quality Control

As stated for biomarkers, any clinical pathology test, assay kit or instrument must be validated and assessed for acceptable quality performance by regular standard procedures. This includes validation of all newly acquired laboratory equipment, validation of instruments that have been moved or undergone maintenance procedures and assessment of the interchangeability of results generated by similar types of instrument, either in the same laboratory or in companion ones. A number of articles and recommendations have been published in recent years that provide detailed information concerning minimal standards (Flatland et al., 2014; Jensen and Kjelgaard-Hansen, 2006). A minimal database should include studies on accuracy, precision and linearity under dilution, as well as spiking experiments to assess potential matrix (i.e. blood properties in a particular species) and interference effects (i.e., haemolysis, lipaemia and icterus). An important component of validation efforts is leftover specimens that are not needed as backup material in archived studies. The use of such specimens for validation studies must also be mentioned in the study protocols, as such data will not fall under a particular study objective, and it requires an efficient and reliable freezer management.

In addition, daily calibration and quality-control procedures should be standard with any automated or manual testing procedure (Camus et al., 2015). The participation in assays, usually on a monthly basis, with external quality-control materials, is highly recommended (Camus et al., 2015).

Finally, the FDA (2015) requires the validation of computer transfer and archiving processes. Validation studies essentially verify the correct real-time data transfer from an automated analyser to the LIMS. Archived data, by definition, may not be altered in any way; however, if modifications do occur, they must be traceable and well documented.

6.11 Data Analysis and Interpretation

Clinical pathology data analysis appears very straightforward, since the numerical data can easily be analysed for statistically significant differences between control and treated groups. However, practice and experience teach us to take a different approach in many instances, and a 'weight-of-evidence' approach is recommended when examining test article-related effects. In non-rodent species, such as dogs, monkeys and pigs, interanimal variation can be considerable, affecting the finding of statistically relevant differences when comparing the means of treated groups with the control group. It is therefore critical to consider comparison with pretreatment values (Hall and Everds, 2003; Tomlinson et al., 2013). In fact, in non-rodent species, a comparison between pre- and post-treatment data in each individual animal is often more informative, although it is a laborious exercise (Hall and Everds, 2003; Tomlinson et al., 2013). It is recommended that study personnel export data into Excel-type documents in order to facilitate such analyses and avoid calculation errors.

Data analysis should also consider other variables and parameters, such as body weight gain and food consumption, toxicokinetics and any other in-life observations (e.g. menstruation in monkeys, test article-related vomiting or diarrhoea, compromised haemostasis after blood sampling), and integrate them into the interpretation of the clinical pathology data.

The basic questions in toxicologic clinical pathology data assessment are (Figure 6.7):

- Are there any changes?
- If yes, are these changes test article-related?

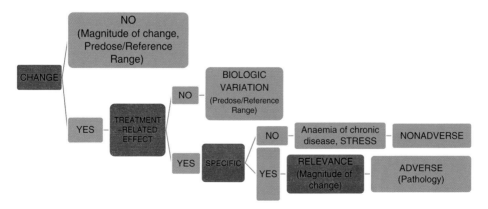

Figure 6.7 Clinical pathology data-interpretation algorithm.

A relation to the test item can be supported by 'yes' answers to the following questions:

- Are the types and magnitudes of the changes comparable in males and females?
- Are the changes dose-related?
- Are the changes reversible (e.g. in studies with a recovery group of animals)?

Finally, an assessment should be made as to:

- whether test article-related changes are toxicologically relevant;
- whether these changes are adverse.

Evidence for test article-related changes can be supported by similar changes in males and females and a clear dose relationship. The relevance and importance of the observed changes are indicated by their magnitude of change, their reversibility and their specificity for a particular treatment. The assessment of adversity in clinical pathology data should not be done in isolation.

6.12   Reporting

It is common that study directors or clinical pathologists are required to report clinical pathology data long before other study results are available. While a preliminary assessment of clinical pathology data is often possible, and can be considered legitimate to some degree, the final report should never be issued without a complete consideration of all other relevant study information (Tomlinson et al., 2013). In particular, results of anatomic pathology – but also in-life observations and the outcomes of toxicokinetics – must be available for the final clinical pathology data interpretation (Kerlin et al., 2016). This applies particularly to situations in which clinical pathology data are included in the determination of no-observed-adverse-effect level (NOAEL) or no-observed-effect level (NOEL) (Kerlin et al., 2016; Ramaiah et al., 2017).

Often, companies require a particular report format. Usually, it seems appropriate to start out with a general statement about whether or not test article-related effects were observed, and to point out some of the major findings. A more detailed narrative listing of the observed changes should follow in the form of system-related profiles (e.g. variables concerning liver function and integrity, or renal function and integrity) or metabolic categories (e.g. glucose metabolism), in order to provide a pathophysiologic framework that allows the best understanding of the mechanisms of potential test article-related effects. Finally, all findings should be considered in terms of biologic and physiologic significance (Figure 6.7) (Tomlinson et al., 2013; Kerlin et al., 2016).

Clearly, specific information on the pharmaceutical actions of the test item in question should also be available to the clinical pathologist when he or she compiles the final report. If the test item is a biological or a small molecule, potential specific information on pharmaceutical targets can be very helpful in interpreting white blood cell data (e.g. with immunmodulating test items) or equivocal liver profiles (e.g. with molecules affecting the hepatobiliary system).

6.13   Food Consumption and Body Weight (Gain)

Recording of food consumption and changes in body weight or body weight gain provide important information on the overall wellbeing of animals in studies (Hoffman et al., 2002) Even though test article-related changes in food consumption and body weight gain may not be specific, such data can provide helpful information on the interpretation of clinical pathology findings. For instance, decreased food consumption can result in a stress response (particularly in rats) and dehydration, followed by changes in the haemogram and chemistry profile, respectively, of affected animals. Typically, subchronic stress-related changes include higher total white blood cell counts, due to higher neutrophil, and lower lymphocyte and eosinophil numbers (Everds et al., 2013). In addition, and particularly in dogs, ALP activity can be induced as a result of increased cortisol levels. In dehydration, the haematocrit is higher ('haemoconcentration'), which can potentially mask a mild loss of red blood cell mass. In addition, higher concentrations of albumin usually result from dehydration. As in haematology, dehydration and haemoconcentration can mask a mild loss of protein and albumin. Finally, a protracted decrease in food consumption can lead to changes in the haematology profile (Asanuma et al., 2010).

6.14   Organ Weights

Organ weights can provide valuable information for the interpretation of clinical pathology results (Sellers et al., 2007). Overall reduced body and organ weights in treated animals must be correlated with food uptake, as they can indicate a massive lack of nutrient intake, which can also be reflected in lower concentrations of total protein and albumin in serum or plasma.

If body weight is lost due to diarrhoea, changes in electrolyte concentrations and overall dehydration and haemoconcentration may also be expected. In rats, glucose concentration and ALP activity will be changed, whereas in vomiting dogs, chloride concentration will decrease and bicarbonate concentration may increase. Other changes, such as generalised lower weights of lymphoid organs (e.g. lymph nodes, thymus and spleen), combined with increased weights of adrenal glands, can support an interpretation of a stress response (Everds et al., 2013). Finally, increased liver weights can be suggestive of test article-related hypertrophy and induction of liver enzymes (Ennulat et al., 2010). Other changes in organ weights (e.g. heart, brain and kidneys) call for thorough histopathologic examination, although increased enzyme activities (e.g. ALT and CK) and troponin I concentrations can assist in assessments of myocardial integrity, while urea and creatinine concentrations in combination with urinalysis provide valuable information on renal function. Increased spleen weights in mice usually indicate increased extramedullary haematopoietic activity.

6.15   Examples of Typical Clinical Pathology Profile Changes in Toxicologic Clinical Pathology

Some common nonspecific change patterns and chemistry-profile patterns are given in Tables 6.6 and 6.7.

Table 6.6 Common nonspecific change patterns in toxicologic clinical pathology.

Variable	Decreased red blood cell mass due to chronic disease	Decreased red blood cell mass due to blood loss or haemolysis	Stress reaction	Inflammatory response
Haematocrit	↓	↓	↔	↓
Haemoglobin concentration	↓	↓	↔	↓
Red blood cell count	↓	↓	↔	↓
Mean cell volume (MCV)	↔	↑	↔	↔
Mean cell haemoglobin (MCH)	↔	↔	↔	↔
Mean cell haemoglobin concentration (MCHC)	↔	↔/↓	↔	↓
Reticulocyte count	↔/↓	↑	↔	↔/↓
Red cell distribution width (RDW)	↔	↑	↔	↔
Total white blood cell count	↔/↑	↑	↑	↑
Neutrophils	↔/↑	↑	↑	↑
Lymphocytes	↔	↑	↓ (↑ peracutely)	↑
Monocytes	↔	↑	↔	↑
Eosinophils	↔	↑	↓	↔/↑
Basophils	↔	↑	↔	↔/↑
Large unclassified cells (LUCs)	↔	↑	↔	↑
Clinical chemistry	↑ Acute phase Proteins, ↓ Albumin, ↓ Calcium, ↔/↑ Globulins	↓ Albumin, ↓ Calcium, ↔/↓ Globulins	↑ Glucose, ↑ ALP (dog)	↓ Albumin, ↑ Globulins
Anatomic pathology	-	Presence of haemorrhage	↓ size/weight Lymph nodes, thymus, spleen	↑ Reticuloendothelial system (RES)

↑, increase; ↔, no change; ↓, decrease.

Table 6.7 Common patterns in chemistry profiles in toxicologic clinical pathology.

Analyte	Dehydration/ haemoconcentration	Renal dysfunction	Hepatic/hepatobiliary dysfunction, disturbed integrity
Aspartate aminotransferase (AST)	↔	↔	↑↔
Alanine aminotransferase (ALT)	↔	↔	↑↔
Sorbitol dehydrogenase (SDH) (minipigs)	↔	↔	↑↔
Glutamate dehydrogenase (GLDH) (minipigs)	↔	↔	↑↔
Alkaline phosphatase (ALP)	↔	↔	↑↔
γ-glutamyltransferase (GGT) (non-rodents)	↔	↔	↑↔
Total bilirubin	↔	↔	↑↔
Glucose	↔	↔	↔
Urea	↑	↑	↓
Creatinine	↑	↑	↔
Triglycerides	↔		↔/↓
Total cholesterol	↔	↑	↔/↑
Total protein	↑	↓	↔/↓
Albumin	↑	↓	↓
Albumin-to-globulin (A : G) ratio	↔/↑	↔/↓	↓
Sodium	↔/↑	↓	↔
Potassium	↔/↑	↑	↔
Chloride	↔/↑	↔/↓	↔
Total calcium	↔	↔/↓	↔
Phosphorus	↔	↔/↑	↔
Haematology	↑ RBC, HGB, HCT	↓ RBC, HGB, HCT	
Coagulation	-	-	↑ PT, APTT
Urinalysis	↑ SG	↔/↓ SG	Bilirubin

↑,increase; ↔, no change; ↓, decrease.

6.15.1 Reduced Red Blood Cell Mass due to Chronic Disease

Reduced red blood cell mass due to chronic disease is a very frequently encountered and reversible test article-related haematologic change in toxicologic studies. It is typically a mild to moderate (up to 15%) loss of red blood cell mass, without an adequate regenerative reticulocyte response. The most likely mechanism is an inhibition of haemoglobin synthesis in immature red blood cells due to the sequestration of iron. This mechanism has been well investigated in clinical medicine in animals and people, and seems to be quite common in preclinical studies, although the exact reason for it is unknown. One important factor is the production of hepcidin in the liver, often elicited by interleukin 6 (IL-6) and other inflammatory mediators (Grimes and Fry, 2015). It can sometimes be helpful to correlate the haematologic finding with markers of an acute- or chronic-phase response, such as fibrinogen, CRP, serum amyloid A or α1-acid glycoprotein concentrations, according to the species, as this indicates that the animal is suffering from chronic inflammation.

6.15.2 Stress Response

Stress reactions in animals are quite common in preclinical studies. A typical subacute or chronic stress response is characterised by increased counts of total leukocytes, usually due to increased counts of neutrophils and possibly monocytes, decreased counts of lymphocytes and possibly eosinophils, increased concentrations of glucose (particularly in rats) and increased activity of ALP (particularly in dogs) (Everds et al., 2013).

6.15.3 Reduced Red Blood Cell Mass due to Excessive Blood Sampling

Not uncommonly, animals – including controls – demonstrate a loss of red-blood-cell mass of up to 20% by the end of a 4-week study. This can be considered a welfare issue and should be avoided. In smaller species, such as monkeys, excessive blood sampling can also cause animal loss, especially in menstruating females (Perigard et al., 2016).

To calculate acceptable blood volumes for sampling, a simple formula that assumes total blood volume is about 10% of the body weight can be used: roughly, no more than 10% of the total blood volume should be removed at any one time (Diehl et al., 2001). An important issue in study animals undergoing excessive blood sampling is the normal physiologic response: namely, the strong stimulation of erythropoiesis in bone marrow and a resulting reticulocytosis in peripheral blood. Even when blood volumes are respected, the blood-sampling regimen may trigger a compensatory erythroid response from the bone marrow, which can easily obscure real test article-related effects in bone-marrow analysis. Therefore, particularly with test items that affect haematopoiesis, oversampling should be avoided in order to prevent artefactual results.

6.15.4 Common Artefacts

An artefact is a study-related effect that produces a change in clinical pathology data related not to the test item, but to circumstantial, procedural or accidental influences. It can include any of the following: haemolysis (at blood sampling), lipaemia (accidental feeding immediately prior to blood sampling), coagulation (at blood sampling), use of incorrect tubes (e.g. K2- or K3-EDTA) and contamination (urine).

Occurrences such as haemolysis, lipaemia and icterus must always be documented, and appropriate measures such as flagging of affected results need to be taken. In addition, the origin of the artefact must be determined. For instance, test item-related haemolysis can be addressed by in vitro testing and feeding-related lipaemia can be avoided by instituting fasting before sampling, while icterus must be followed up by detailed assessment of liver function and integrity, including indirect bilirubin.

Clearly, almost all artefacts can be prevented by well-standardised practices and properly trained personnel.

6.15.4.1 Coagulation

Technical personnel running coagulation assays should be properly trained in the mechanisms and principles of coagulation (namely, the intrinsic and the extrinsic pathways) in order to be capable of recognising and troubleshooting erroneous data. It is important to remember that coagulation in preclinical studies is measured in seconds, representing a completely different pattern of changes related to treatment. A prolongation of coagulation times is most often caused by a lack of coagulation factors, which can result from reduced production (e.g. in the liver, due to hepatic dysfunction) or loss of coagulation factors (when samples clot during collection). The latter can happen particularly in younger, female animals, and requires careful screening for blood clots prior to coagulation assays or, at the latest, when irregularities are observed.

6.15.4.2 Haemolysis and Increased Potassium Concentration

There are variable concentrations of potassium inside red and other blood cells, including platelets. Lysis of red blood cells or activation and aggregation of platelets can result in potassium release and transiently increased concentrations of potassium in serum or plasma. Therefore, the degree of haemolysis should always be recorded, alongside other potential inference from lipaemia and icterus. Whenever possible, haemolysis and lipaemia should be avoided by following proper venipuncture technique and correct handling of collected blood, and by fasting animals prior to sampling their blood. However, lipaemia, haemolysis and icterus can also represent a test article-related effect.

6.15.4.3 Blood in Urine

Urine collection in preclinical studies is usually done with metabolic cages, where voided urine drips down through the cage floor into a collection pan over an extended time period of several hours. As a recent study showed, contamination with food or faeces can result in a considerable number of false-positive blood results in such urine samples from dogs and monkeys (Aulbach et al., 2015a). Therefore, it is recommended that urine collection procedures be optimised to avoid misleading results falsely implying potentially serious test article-related changes in the urinalysis.

6.16 Microsampling

Microsampling is currently being discussed in many spheres, mainly with the purpose of reducing the blood volumes necessary for analysis, and thus the number of animals needed in a given study. This applies mainly to mice and rats, but also to dogs and monkeys.

Although a few available studies may suggest that simple sample dilution yields acceptable results, others raise concerns over inaccurate and imprecise results (Zamfir et al., 2014; Aulbach et al., 2015b). It is recommended that regular validation studies be performed and that a minimal database be kept recording accuracy, precision, linear dilution effects and determination of proportional and constant bias in comparison to standard methods, in order to ensure the equivalency of microsampling to established standard methods (Zamfir et al., 2014; Moorhead et al., 2016; Poitout-Belissent et al., 2016).

6.17 Conclusion

Clinical pathology is a well-established discipline in preclinical sciences. Although data generation and assessment appear relatively straightforward, based on the mostly numerical quantitative changes, clinical pathology also includes qualitative/morphological data evaluation, which requires specialised training. Many considerations affect data interpretation and the determination of test article-related changes. Therefore, it is advantageous if those individuals active in all aspects of clinical pathology data generation are properly trained and understand current best practice.

Acknowledgments

My sincere thanks go to Dr Ilse Schwendenwein for providing some of the images, and Drs Anne Provencher and Michael Goe for their critical review of the manuscript.

References

Abbas, A.K. and Lichtman, A.H.H. (2014) Cellular and Molecular Immunology, 8th edn, Elsevier, Amsterdam.

Ameri, M., Schnaars, H.A., Sibley, J.R. and Honor, D.J. (2011) Determination of plasma fibrinogen concentrations in beagle dogs, cynomolgus monkeys, New Zealand white rabbits, and Sprague–Dawley rats by using Clauss and prothrombin-time-derived assays. *Journal of the American Association for Laboratory Animal Science*, 50(6), 864–7.

Asanuma, F., Miyata, H., Iwaki, Y. and Kimura, M. (2010) Feature on erythropoiesis in dietary restricted rats. *Journal of Veterinary Medical Science*, 73(1), 89–96.

Aulbach, A.D., Schultze, E., Tripathi, N.K., Hall, R.L., Logan, M.R. and Meyer, D.J. (2015a) Factors affecting urine reagent strip blood results in dogs and nonhuman primates and interpretation of urinalysis in preclinical toxicology studies: a Multi-Institution Contract Research Organization and BioPharmaceutical Company Perspective. *Veterinary Clinical Pathology*, 44, 229–33.

Aulbach, A., Ramaiah, L., Tripathi, N. and Poitout, F. (2015b) Industry survey results on clinical pathology volume requirements in preclinical toxicological studies. Poster presented at the Society of Toxicology.

Bennett, J.S., Gossett, K.A., McCarthy, M.P. and Simpson, E.D. (2009) Effects of ketamine hydrochloride on serum biochemical and hematologic variables in rhesus monkeys (Macaca mulatta). *Veterinary Clinical Pathology*, 21, 15–18.

Braun, J.P., Concordet, D., Geffré, A., Bourges Abella, N. and Trumel, C. (2013) Confidence intervals of reference limits in small reference sample groups. *Veterinary Clinical Pathology*, 42, 395–8.

Braun, J.P., Bourgès-Abella, N., Geffré, A., Concordet, D. and Trumel, C. (2015) The preanalytical phase in veterinary clinical pathology. *Veterinary Clinical Pathology*, 44(1), 8–25.

Burkhardt, J.E., Pandher, K., Solter, P.F., Troth, S.P., Boyce, R.W., Zabka, T.S. and Ennulat, D. (2011) Recommendations for the evaluation of pathology data in nonclinical safety biomarker qualification studies. *Toxicologic Pathology*, 39, 1129–37.

Camus, M.S., Flatland, B., Freeman, K.P. and Cruz Cardona, J.A. (2015) ASVCP quality assurance guidelines: external quality assessment and comparative testing for reference and in-clinic laboratories. *Veterinary Clinical Pathology*, 44(4), 477–92.

Christensen, M.B., Langhorn, R., Goddard, A., Andreasen, E.B., Moldal, E., Tvarijonaviciute, A., Kirpensteijn, J., Jakobsen, S., Persson, F. and Kjelgaard-Hansen, M. (2014) Comparison of serum amyloid A and C-reactive protein as diagnostic markers of systemic inflammation in dogs. *Canadian Veterinary Journal*, 55, 161–8.

Cray, C., Zaias, J. and Altman, N.H. (2009) Acute phase response in animals: a review. *Comparative Medicine*, 59(6), 517–26.

Diehl, K.H., Hull, R., Morton, D., Pfister, R., Rabemampianina, Y., Smith, D., Vidal, J.M. and van de Vorstenbosch, C.; European Federation of Pharmaceutical Industries Association and European Centre for the Validation of Alternative Methods. (2001) A good practice guide to the administration of substances and removal of blood, including routes and volumes. *Journal of Applied Toxicology*, 21, 15–23.

Ennulat, D., Walker, D., Clemo, F., Magid-Slav, M., Ledieu, D., Graham, M., Botts, S. and Boone, L. (2010) Effects of hepatic drug-metabolizing enzyme induction on clinical pathology parameters in animals and man. *Toxicologic Pathology*, 38(5), 810–28.

Everds, N.E., Snyder, P.W., Bailey, K.L., Bolon, B., Creasy, D.M., Foley, G.L., Rosol, T.J. and Sellers, T. (2013) Interpreting stress responses during routine toxicity studies: a review of the biology, impact, and assessment. *Toxicologic Pathology*, 41(4), 560–614.

FDA. (2015) Code of Federal Regulations (CFR) Title 21. Part 58 Good Laboratory Practice for nonclinical laboratory studies. US Food and Drug Administration. Available from: http://www.accessdata.fda.gov/scripts/cdrh/cfdocs/cfcfr/CFRSearch.cfm?CFRPart=58&showFR (last accessed July 29, 2016).

Flatland, B., Friedrichs, K.R. and Klenner, S. (2014) Differentiating between analytical and diagnostic performance evaluation with a focus on the method comparison study and identification of bias. *Veterinary Clinical Pathology*, 43, 475–86.

Friedrichs, K.R., Harr, K.E., Freeman, K.P., Szladovits, B., Walton, R.M., Barnhart, K.F. and Blanco-Chavez, J.; American Society for Veterinary Clinical Pathology. (2012) ASVCP reference interval guidelines: determination of de novo reference intervals in veterinary species and other related topics. *Veterinary Clinical Pathology*, 41, 441–53.

Geffre, A., Braun, J.P., Trumel, C. and Concordet, D. (2009) Estimation of reference intervals from small samples: an example using canine plasma creatinine. *Veterinary Clinical Pathology*, 38, 477–84.

Geffre, A., Concordet, D., Braun, J.P., Trumel, C. (2011) Reference Value Advisor: a new freeware set of macroinstructions to calculate reference intervals with Microsoft Excel. *Veterinary Clinical Pathology*, 40, 107–12.

Grimes, C.N. and Fry, M.M. (2015) Nonregenerative anemia: mechanisms of decreased or ineffective erythropoiesis. *Veterinary Pathology*, 52(2), 298–311.

Gunn-Christie, R.G., Flatland, B., Friedrichs, K.R., Szladovits, B., Harr, K.E., Ruotsalo, K., Knoll, J.S., Wamsley, H.L. and Freeman, K.P.; American Society for Veterinary Clinical Pathology (ASVCP). (2012) ASVCP quality assurance guidelines: control of preanalytical, analytical, and postanalytical factors for urinalysis, cytology, and clinical chemistry in veterinary laboratories. *Veterinary Clinical Pathology*, 41, 18–26.

Hall, R.L. and Everds, N.E. (2003) Factors affecting the interpretation of canine and nonhuman primate clinical pathology. *Toxicologic Pathology*, 31, 6–10.

Heegaard, P.M.H., Miller, I., Sorensen, N.S.S., Soerensen, K.E. and Skovgaard, K. (2013) Pig α1-acid glycoprotein: characterization and first description in any species as a negative acute phase protein. *PLoS One*, 8(7), e68110.

Hoffman, M. and Monroe, D.M. III. (2001) A cell-based model of hemostasis. *Thrombosis and Haemostasis*, 85, 958–65.

Hoffmann, W.P., Ness, D.K. and Van Lier, R.B.L. (2002) Analysis of rodent growth data in toxicology. *Toxicological Sciences*, 66, 313–19.

Honjo, T., Kuribayashi, T., Matsumoto, M., Yamazaki, S. and Yamamoto, S. (2010) Kinetics of α2-macroglobulin and α1-acid glycoprotein in rats subjected to repeated acute inflammatory stimulation. *Laboratory Animals*, 44(2), 150–4.

Jensen, A.L. and Kjelgaard-Hansen, M. (2006) Method comparison in the clinical laboratory. *Veterinary Clinical Pathology*, 35, 276–86.

Jordan, H.L., Register, T.C., Tripathi, N.K., Bolliger, A.P., Everds, N., Zelmanovic, D., Poitout, F., Bounous, D.I., Wescott, D. and Ramaiah, S.K. (2014) Nontraditional applications in clinical pathology. *Toxicologic Pathology*, 42(7), 1058–68.

Kaneko, J., Harvey, J. and Bruss, M. (2008) Clinical Biochemistry of Domestic Animals, 6th edn, John Wiley & Sons, Chichester.

Kerlin, R., Bolon, B., Burkhardt, J., Francke, S., Greaves, P., Meador, V. and Popp, J. (2016) Scientific and Regulatory Policy Committee: recommended ('best') practices for determining, communicating, and using adverse effect data from nonclinical studies. *Toxicologic Pathology*, 44(2), 147–62.

Kurata, M. and Horii, I. (2004) Blood coagulation tests in toxicological studies, review of methods and their significance for drug safety assessment. *Journal of Toxicological Sciences*, 29(1), 13–32.

Moorhead, K.A., Discipulo, M.L., Hu, J., Moorhead, R.C. and Johns, JL. (2016) Alterations due to dilution and anticoagulant effects in hematologic analysis of rodent blood samples on the Sysmex XT-2000iV. *Veterinary Clinical Pathology*, 45(2), 215–24.

NC3Rs. (2016) Blood sampling. National Centre for the Replacement, Refinement & Reduction of Animals in Research. Available from: https://www.nc3rs.org.uk/our-resources/blood-sampling (last accessed July 29, 2016).

Perigard, C.J., Parrula, M.C., Larkin, M.H. and Gleason, CR. (2016) Impact of menstruation on select hematology and clinical chemistry variables in cynomolgus macaques. *Veterinary Clinical Pathology*, 45(2), 232–43.

Poitout-Belissent, F., Aulbach, A., Tripathi, N. and Ramaiah, L. (2016) Reducing blood volume requirements for clinical pathology testing in toxicologic studies - Points to consider. *Veterinary Clinical Pathology*. In press.

Ramaiah, L., Tomlinson, L., Tripathi, N., Cregar, L., Vitsky, A., von Beust., Barbara., Barlow, V., Reagan, W. and Ennulat, D. Principles for assessing adversity in toxicologic clinical pathology – Points to consider. In Press.

Reagan, W.J., Irizarry-Roviram, A., Poitout-Belissent, F., Bollinger, A.P., Ramaiah, S.K., Travlos, G., Walker, D., Bounous, D. and Walter, G.; Bone Marrow Working Group of ASVCP/STP. (2011) Best practices for evaluation of bone marrow in nonclinical toxicity studies. *Toxicologic Pathology*, 39(2), 435–48.

Rütgen, B.C., König, R., Hammer, S.E., Groiss, S., Saalmüller, A. and Schwendenwein, I. (2015) Composition of lymphocyte subpopulations in normal canine lymph nodes. *Veterinary Clinical Pathology*, 44(1), 58–69.

Sellers, R.S., Morton, D., Michael, B., Roome, N., Johnson, J.K., Yano, B.L., Perry, R. and Schafer, K. (2007) Society of toxicologic pathology position paper: organ weight recommendations for toxicology studies. *Toxicologic Pathology*, 35, 751–5.

Tomlinson, L., Boone, L.I., Ramaiah, L., Penraat, K.A., von Beust, B.R., Ameri, M., Poitout-Belissent, F.M., Weingand, K., Workman, H.C., Aulbach, A.D., Meyer, D.J., Brown, D.E., MacNeill, A.L., Bolliger, A.P. and Bounous, D.I. (2013) Best practices for veterinary toxicologic clinical pathology, with emphasis on the pharmaceutical and biotechnology industries. *Veterinary Clinical Pathology*, 42, 252–69.

Vap, L.M., Harr, K.E., Arnold, J.E., Freeman, K.P., Getzy, K., Lester, S. and Friedrichs, K.R.; American Society for Veterinary Clinical Pathology (ASVCP). (2012) ASVCP quality assurance guidelines: control of preanalytical and analytical factors for hematology for mammalian and nonmammalian species, hemostasis, and crossmatching in veterinary laboratories. *Veterinary Clinical Pathology*, 41, 8–17.

Weiss, D.J. and Wardrop, K.J. (2010) Schalm's Veterinary Hematology, 6th edn, Wiley-Blackwell, Boston, MA.

Zamfir, M., Prefontaine, A., Copeman, C., Poitout, F. and Provencher, A. (2014) Dilution of blood for hematology and biochemistry analysis in rats. Poster presentation, ASVCP Annual Conference 2014, Atlanta, GA.

7

Adversity: A Pathologist's Perspective

Bhanu Singh

Discovery Sciences, Janssen Research & Development, Spring House, PA, USA

Learning Objectives

- Define NOEL, NOAEL and LOAEL.
- Define adversity.
- Consider relevant factors when determining the adversity of pathology findings.
- Communicate NOAEL in toxicity studies.

Worldwide regulatory agencies require data from toxicologic studies in animals in order to govern the introduction of drugs, consumer products, and food additives to the human population, as well as chemical exposure in the environment. Toxic effects in these animal studies are generally expected to occur in a dose-dependent manner in relation to their incidence and severity, allowing the determination of dose levels where substantial effects occur or where these effects do not occur. Toxic effects generally manifest as changes in cell or tissue morphology and/or function. Gross, microscopic and clinical pathological evaluation are central to toxicological assessment in these studies, and often help in deciding which doses produce an adverse or a nonadverse effect.

Before determining 'adversity', a pathologist must address various questions and make decisions at various levels (Figure 7.1). At an individual animal level, a finding in a tissue or organ has to be judged as noteworthy and must be differentiated from artefacts caused by post mortem change or tissue processing. The pathologist has to give the finding an appropriate and a globally accepted and harmonised term. At the study level, the pathological findings must be judged in relation to the test article. To differentiate a test article effect from a chance finding, consideration should be given to the dose response, the ranges of natural variation and the biological plausibility (e.g. clinical/nonclinical data for other compounds in same class, mode of action) (ECETOC, 2000; Lewis et al., 2002).

Often the terms 'treatment' and 'test article' are used interchangeably when describing an effect. It is important to distinguish whether a finding is 'treatment'- or 'test article'-related. 'Treatment' indicates the effects likely to be attributed to a procedure (e.g. haemorrhages around the eye due to retro-orbital bleeding). In a study designed with appropriate controls, procedure- or vehicle-related changes are expected to occur

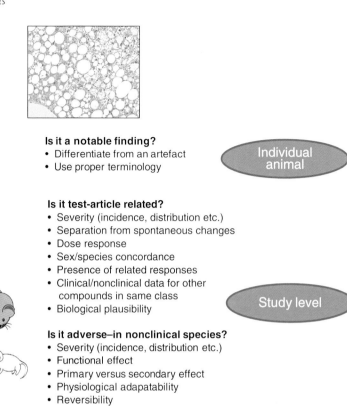

Is it a notable finding?
- Differentiate from an artefact
- Use proper terminology

Individual animal

Is it test-article related?
- Severity (incidence, distribution etc.)
- Separation from spontaneous changes
- Dose response
- Sex/species concordance
- Presence of related responses
- Clinical/nonclinical data for other compounds in same class
- Biological plausibility

Study level

Is it adverse–in nonclinical species?
- Severity (incidence, distribution etc.)
- Functional effect
- Primary versus secondary effect
- Physiological adapatability
- Reversibility
- Pharamcological effect

Is it adverse–in humans?
- Translation ability of animal finding
- Human exposure of chemical

Human health risk assessment

Figure 7.1 Approach to defining the adversity of a pathological finding.

in all dose groups, but in studies with a small number of animals or no control animals (exploratory studies), these findings may not be distributed evenly in all the groups and can complicate the interpretation.

This chapter provides a pathologist's perspective on how to determine the adversity of pathological findings in toxicity studies. It will also discuss the definitions of various toxicity terms, such as 'lowest observable adverse effect level' (LOAEL), 'lowest observable effect level' (LOEL), 'no observable adverse effect level' (NOAEL) and 'no observable effect level' (NOEL), along with challenges in defining adversity and communication of NOAEL in study reports.

7.1 LOAEL, NOEL and NOAEL: Definition

As part of a risk assessment, toxicology studies in animals are conducted in order to identify and characterise the toxic effects of chemicals and drugs (hazard identification).

Based on the dose response in a toxicological study, various dose-related indices (LOAEL, NOAEL, NOEL) commonly used in risk assessment can be calculated. In a well-designed study, doses are selected which produce a clear toxic effect, a LOAEL and either a LOEL, a NOAEL or a NOEL.

As defined in various toxicology publications (e.g. Klaassen, 2013; Derelanko and Auletta, 2014):

- LOAEL is the lowest dose at which there are statistically and/or biologically significant increases in the frequency or severity of adverse effects between the exposed population and its appropriate control.
- NOAEL is the highest experimental dose at which there are no statistically and/or biologically significant increases in the frequency or severity of adverse effects between the exposed population and its appropriate control.
- NOEL is the highest experimental dose at which there are no test article-related effects (adverse or nonadverse) observed in the exposed population, when compared with its appropriate control.

It is noteworthy that the NOAEL is a measured or estimated value and may be different from the true no-adverse-effect level, which lies somewhere between the measured NOAEL and the measured LOAEL (Lewis et al., 2002; Filipsson et al., 2003). Once the NOAEL or NOEL is established from animal toxicity studies, regulatory guidelines are harmonised around the calculation of a safe starting dose for human clinical trials or the determination of allowable exposure limits (FDA, 2005; ICH, 2009; EPA, 2012).

7.2 Adversity

It is a well accepted fact amongst toxioclogists and pathologists that there is no consensus on the definition of 'adversity' or 'adverse effect', or even of 'NOAEL'. This can be seen from a quick review of the published literature and various regulatory guidelines (Table 7.1). In general, however, all definitions indicate that an adverse effect is a change in biochemical, functional or structural parameters that may impair performance and generally have a harmful effect on the growth, development or life span of a nonclinical toxicology model (Dorato and Engelhardt, 2005).

Much controversy exists regarding the relevance of extrapolating toxicity findings from nonclinical studies to a clinical setting. According to some definitions, an adverse finding is an adverse finding regardless of its relevance to humans. The US Food and Drug Administration (FDA) definition states that 'the use of NOAEL as a benchmark for dose-setting in healthy volunteers should be acceptable to all responsible investigators. As a general rule, an adverse effect observed in nonclinical toxicology studies used to define a NOAEL for the purpose of dose-setting should be based on an effect that would be unacceptable if produced by the initial dose of a therapeutic in a phase 1 clinical trial conducted in adult healthy volunteers' (FDA, 2005). Note that the FDA does not ask that the pathologist or toxicologist determine whether the adverse effect would occur in humans or comment in any way on the relevance of the finding for humans; it simply asks, if it occurred, would it be acceptable for a healthy volunteer who can derive no therapeutic benefit from the drug?

Table 7.1 Selected definitions of 'adverse effect' from the published literature.

Source	Definition
EPA (2007)	'Adverse effect: A biochemical change, functional impairment, or pathological lesion that affects the performance of the whole organism, or reduces an organism's ability to respond to an additional environmental challenge.'
FDA (2005)	'...an adverse effect observed in nonclinical toxicology studies used to define a NOAEL for the purpose of dose-setting should be based on an effect that would be unacceptable if produced by the initial dose of a therapeutic in a phase 1 clinical trial conducted in adult healthy volunteers.'
IPCS (2004)	'...change in the morphology, physiology, growth, development, reproduction, or life span of an organism, system, or (sub) population that results in an impairment of functional capacity, an impairment of the capacity to compensate for additional stress, or an increase in susceptibility to other influences.'
Sergeant (2002)	'Adverse effects are changes that are undesirable because they alter valued structural or functional attributes of the entities of interest...The nature and intensity of effects help distinguish adverse changes from normal...variability or those resulting in little or no significant change.'
Lewis et al. (2002)	'...biochemical, morphological or physiological change (in response to a stimulus) that either singly or in combination adversely affects the performance of the whole organism or reduces the organism's ability to respond to an additional environmental challenge.'
Eaton and Gilbert (2008)	'The spectrum of undesired effects of chemicals is broad. Some effects are deleterious and others are not...[Regarding drugs], some side effects...are never desirable and are deleterious to the well-being of humans. These are referred to as the adverse, deleterious, or toxic effects of the drug.'

The difference in approach likely arises from the nature of chemicals and from differences between the agencies responsible for making regulatory decisions. For example, risk assessment performed by the US Environmental Protection Agency (EPA) for the registration and marketing of agrochemicals relies heavily on animal toxicity data, with little or no information on human exposure. Environmental chemicals are not intended to provide therapeutic benefits. In contrast, for a pharmaceutical drug, critical factors in the registration and marketing are human efficacy and safety data. In early drug development, due to lack of human-exposure data, animal-toxicity data generally help in making 'go or no-go' efficacy and safety decisions; later in development, toxicity data can still be decisive, especially regarding reproductive/developmental toxicity or carcinogenic potential.

With the growth of advanced technologies and high-throughput approaches in the field of toxicity testing, there is a strong interest in finding alternatives to animal testing (NRC, 2007). At the centre of this strategy is a revamping of testing to focus on the molecular mechanisms of toxicant effects. Using high-throughput assays and human cells in vitro, large amounts of in vitro data can be generated. In order to make biological-interpretation and regulatory decisions based on these data, it is critical to re-evaluate the criteria of adversity that consider the toxicant-induced mode of action at the molecular and cellular levels. A Health and Environmental Sciences Institute (HESI) committee was tasked with discussing approaches to identifying adverse effects in the context of 21st-century toxicity testing. The committee (Keller et al., 2012) recently published what appears to be a practical definition of adversity in this context, adopted from the IPCS risk-assessment terminology (IPCS, 2004), defining an adverse effect as: 'A change in morphology, physiology, growth, development, reproduction, or life span of a cell or organism, system, or (sub)-population that results in an impairment of functional capacity, an impairment of the capacity to compensate for additional stress, or an increase in susceptibility to other influences.'

7.3 Determining Adversity using Pathology Findings: Factors to Consider

Toxicologists and pathologists have not been consistent in applying judgements in determining whether an observed effect in a nonclinical toxicity study is adverse or nonadverse. A nonadverse finding may be given in the following categories: 'no significant effect on organ function', 'no significant effect in overall health', 'finding does not result in organ failure', 'finding being reversible or nonprogressive' 'finding being an adaptive response' and 'finding has no counterpart in humans or occurs through a mechanism not relevant to humans'. Thus, the decision over the adverse nature of a test article-related finding in a nonclinical toxicology study is often subject to discussion, challenge and reinterpretation (Dorato and Engelhardt, 2005).

Determination of a test article-related gross, microscopic or clinical pathology finding as 'adverse' is a challenging task. The major issue concerns the lack of a harmonised definition of 'adverse finding' or of criteria for determining adversity. A consistent approach to defining NOAEL and interpreting adversity in toxicology study reports would be helpful in developing a common understanding of toxicity findings for human health risk assessment. Several efforts have been made recently by professional societies to discuss this important issue and to decide on recommendations to help toxicologists and pathologists (Kerlin et al., 2015).

Recently, the Society of Toxicologic Pathology (STP) has suggested recommendations and best practices for adversity determination (Kerlin et al., 2015), including that adverse findings in study reports pertain only to the test species used within the study, without speculation over potential human relevance or clinical indication. All test article-related adverse findings should be assessed on their own merit and not dismissed due to presumptive pathogeneses.

The collective judgement and experience of the pathologist is the key to determining whether an observed pathological change is adverse or nonadverse. A weight-of-evidence approach that takes into account multiple factors, as described in this section, can be used to determine the adversity of a test article-related pathological finding (Lewis et al., 2002; Dorato and Engelhardt, 2005).

7.3.1 Severity

Severity criteria can be useful to distinguish between adverse and nonadverse effects. A test article-related effect is less likely to be nonadverse if the severity of the change is minimal and is not associated with other adverse findings. This is especially relevant for pathological parameters that are measured objectively (e.g. haematology, clinical chemistry, organ weights) and have defined criteria for 'adversity' (as explained in the example in the next paragraph). Changes in such parameters above a certain level (threshold of concern) will be considered 'adverse'.

Inhibition of acetylcholinesterase in the tissues of the nervous system is generally accepted as a key event in the mechanism of toxicity leading to adverse cholinergic effects. Since it may not be practical to measure this enzyme in the central and peripheral nervous tissues, blood cholinesterase is considered an appropriate surrogate measure of the potential effects on peripheral-nervous-system acetylcholinesterase activity in animals. A threshold (e.g. 20% inhibition) is used to determine a biologically significant depression (adverse effect) of brain and erythrocyte cholinesterase. A statistically significant reduction in these enzyme activities above this threshold is considered to represent an adverse effect. The criterion used to determine the threshold is based largely on a review of toxicity studies containing cholinesterase-inhibiting substances (EPA, 2000).

Severity grading of non-neoplastic lesions is considered 'semiquantitative' because it is based on estimates of severity rather than actual measurements. For each lesion, the severity grade is determined mainly by the extent or an estimate of the percentage of tissue involvement, as well as the magnitude of various components of the lesion. The nomenclature used for severity grading differs between different laboratories according to the computer program used to acquire the data and the preferences of the pathologist. In general, the most commonly used grading schemes utilise four or five severity grades, to which descriptive terms (minimal, mild, moderate, marked etc.) and/or numerical levels (grade 1, 2, 3, 4 etc.) are applied (Shackelford et al., 2002).

In contrast to the example of cholinesterase, defining a threshold of concern for histology (based on degree of severity) is quite challenging. As an example, if 'minimal' changes (e.g. vacuolated macrophages without associated degenerative or necrotic changes in lung parenchyma) are present in the lung during a 1-month study, this can be judged as 'nonadverse' (Nikula et al., 2013). Alternatively, a pathologist might be asked to determine the significance of mild gastric mucosal atrophy following a 1-month repeat-dose toxicity study in rats. Should this finding be considered adverse or nonadverse? The answer to this question will often be based on the training and experience of the pathologist, as well as information available from the scientific literature. Finally, minimal or subclinical changes in a few pathological parameters might be interpreted as adverse findings. Minimal increases in cardiac troponin levels, for example, may be considered adverse (consistent with myocardial degeneration or necrosis) even if there is no morphological evidence of cardiomyocyte damage.

7.3.2 Functional Effect

Functional effect is one well-accepted method of determining adversity. If an effect does not result in alteration in the general function of an organ/tissue or the test organism, it is likely to be nonadverse (Lewis et al., 2002). It is not uncommon to see test article-related effects on multiple anatomical and clinical pathology end points in a toxicity study. In some organ systems (e.g. the nervous system), adverse functional effects can occur even without significant changes in histopathology or clinical pathology parameters. In other organ systems, microscopic changes of lesser severity (e.g. hepatocellular hypertrophy) or clinical pathology parameters (minor changes in red cell mass) generally do not result in functional impairment and are therefore considered nonadverse.

Drug-induced phospholipidosis is an excessive, reversible accumulation of phospholipid and associated drug in lysosomes (Reasor and Kacew, 2001). Phospholipidosis is most often seen in lung, liver and kidneys in nonclinical studies. When it is associated with concurrent inflammatory and/or degenerative changes in tissues, it is considered adverse. Phospholipidosis is commonly seen as accumulation of vacuolated (foamy) macrophages in the lungs, without evidence of tissue injury. To determine the adversity of this finding, it is important to understand the biological consequences of phospholipid accumulations in alveolar macrophages. The effect of amiodarone-induced pulmonary phospholipidosis on pulmonary host-defence functions has been studied in rats. Following one week of daily administration of amiodarone, a 4.5-fold increase occurred in total phospholipid in alveolar macrophages. Phospholipidosis did not impair the immune function of alveolar macrophage phagocytosis, cytokine production (interleukin (IL)-1, IL- 6, tumour necrosis factor alpha (TNF-α)) or pulmonary clearance of *Listeria monocytogenes* – following intratracheal administration of the bacteria (Reasor et al., 1996). Based on these findings, induction of pulmonary phospholipidosis by amiodarone in rats is considered nonadverse.

Even if an effect is regarded as adaptive, it is not necessarily a nonadverse effect, because of its impact on organ function. For example, an adverse effect can result from chronic adaptive change in the tissue. Squamous metaplasia (change of respiratory epithelium to a squamous epithelium) is a commonly seen, microscopic change in the larynx in response to inhaled substances. Initially, the minor epithelial changes of the larynx are characterised as damaged cilia, loss of cilia and flattening of the epithelium. Minor changes in the laryngeal epithelium are graded as minimal and considered nonadverse, because they are not associated with any dysfunction of the larynx. These findings generally progress to multifocal or diffuse laryngeal squamous metaplasia, in which ciliated or columnar laryngeal respiratory epithelial cells are replaced by stratified squamous epithelium (with or without keratinisation), likely resulting in dysfunction of the larynx; this should be considered adverse (Kaufmann et al., 2009).

7.3.3 Primary versus Secondary Effects

In the pathology report, pathologists generally discuss whether the test article-related finding is a primary target effect or is secondary to some other toxicity (e.g. systemic toxicity). The Society of Toxicological Pathology (STP) recommends that determination of the 'adversity' of test article-related changes should be independent of their speculative designations as either primary or subordinate (secondary or tertiary) effects (Kerlin et al., 2015).

A common example of secondary effects in toxicology studies is changes in the organ weights associated with changes in body weight. Understanding the biological response of an organ to body-weight changes can be helpful in differentiating primary versus secondary effects. The organ weights of brain and testes are relatively unaffected by body-weight loss, so in the case of such a loss, the organ weight/body weight ratio of these organs will show an increase. Studies indicate that organ-to-brain weight ratios are a more appropriate parameter for evaluating organ toxicity in the adrenal glands and ovaries, whilst organ-to-body weight ratios are more appropriate for liver and thyroid-gland weights (Bailey et al., 2004).

Animal stress is an integral feature of toxicity studies, mainly due to the use of high doses of test article and/or experimental procedures. Stress-induced changes in the neuroendocrine hormones can result in the alteration of the following end points in toxicity studies: total body weights or body weight gain; food consumption and activity; organ weights (e.g. thymus, spleen, adrenal); lymphocytes in thymus and spleen; circulating leukocyte counts (e.g. increased neutrophils and decreased lymphocytes and eosinophils); and reproductive functions (Everds et al., 2013). Stress responses should be interpreted as secondary (indirect), rather than primary (direct), test article-related findings.

7.3.4 Physiological Adaptability

Normal cells tend to maintain their intracellular environment within a fairly narrow range of physiologic parameters, maintaining a steady state termed 'homeostasis'. As cells are challenged with physiologic stresses or pathologic stimuli (e.g. chemical-induced toxicity), they try to undergo adaptation, achieving a new steady state and maintaining their viability and function. These adaptive responses are changes in the cell size (atrophy or hypertrophy), cell numbers (hyperplasia), phenotype (metaplasia), metabolic activity or cell function. If the adaptive capacity of a cell is exceeded, or if the external stimulus is inherently toxic, cell injury develops. Within certain limits, injury is reversible, and cells return to a stable baseline after removal of stimuli; however, severe or persistent toxicity results in the irreversible injury and death of the affected cells.

Hepatocellular hypertrophy is a commonly observed adaptive change in response to chemicals that are metabolised in the liver before elimination. Phenobarbital is a classical example of such a chemical. In order to increase its functional capability, the hepatocytes increase the synthesis of metabolising enzymes by proliferating smooth endoplasmic reticulum (SER) to eliminate the xenobiotic, resulting in increased cellular size (hypertrophy). As the cellular demand for metabolising enzymes increases, the liver weight and size continue to increase due to the hypertrophy. In this case, hepatocellular hypertrophy without histologic or clinical pathology alterations indicative of liver injury should be considered an adaptive and nonadverse reaction. In cases of adverse reaction, there is a considerable increase in liver size, and single or clustered of necrotic hepatocytes can be seen in subcapsular regions due to the mechanical effects of compression (Parker and Gibson, 1995). An adaptive mechanism (enzyme induction) may become saturated at a certain point (threshold), resulting in direct hepatocellular injury by the chemical or its metabolite. Hepatocellular injury in these cases is generally associated with a significant increase in liver transaminases (alanine aminotransferase (ALT), aspartate aminotransferase (AST)) in the serum and should be considered adverse.

7.3.5 Reversibility of the Lesion

The reversibility of an effect can be used as an important part of the weight of evidence in determining adversity (Lewis et al., 2002). Test article-related effects with partial or full reversibility are less likely to be adverse and can be extrapolated to humans to suggest that a particular finding – if it were to occur in humans – might be transient. To evaluate the reversibility of a toxicity finding, an additional small number of animals must be added to at least the high-dose and control groups of Good Laboratory Practice (GLP) toxicity studies. The duration of the recovery phase is based on the length of the dosing phase of the study, the nature of the changes seen previously and the pharmacokinetic properties of the drug.

The STP has provided recommendations and best practices regarding the need for recovery studies, including the recovery phase in toxicity studies, and has discussed the need to predict recovery in the absence of recovery data (Perry et al., 2013). Pathologists, by virtue of their extensive training and experience, can also provide a reasonable estimate of the likelihood of reversibility in many cases, by taking into account the regenerative capacity of the affected organ system, the effect on the associated extracellular matrix and the pathogenesis of the lesion. If the extracellular matrix is unaffected and the tissue injury involves labile (e.g. bone marrow, haematopoietic tissues, surface epithelium) or stable (e.g. kidney, liver) tissues, recovery is likely. In contrast, if the tissue injury involves permanent tissue (e.g. neurons, cardiac muscle) or results in damage to or loss of the extracellular matrix, full recovery is unlikely, and the more likely outcome is disorderly nonfunctional resolution (e.g. fibrosis) (Perry et al., 2013).

7.3.6 Pharmacological Effect

Toxicity effects in nonclinical toxicity studies are broadly categorised into three mechanisms: chemical-based, on-target (or mechanism-based) and off-target effects. On-target effects are the exaggerated and pharmacologic effects on the target of interest in the test system. Understanding the mechanism of toxicity is important for both risk assessment and the programme development strategy. Since the desirable efficacy also results from modulation of the target, it is often impossible to avoid toxicity due to on-target-related effects without avoiding the target.

  An effect due to a known pharmacological response does not necessarily rule out its adverse nature. Effects caused by exaggerated pharmacology should not be considered adverse as long as there are no indications for organ/tissue damage and/or they are not affecting the general condition of the animal. If a toxic effect as a result of exaggerated pharmacology is considered adverse in a test species, this should be discussed in the study report. Pathologists can provide an explanation as to why the adverse effect is an exaggerated pharmacological effect.

  An example of exaggerated pharmacology can be seen at higher doses in animal studies of erythropoiesis-stimulating agents such as erythropoietins (EMA, 2004). In an anaemic animal or human, an erythropoietin-induced increase in red cell mass can be beneficial; however, findings in toxicity studies (normal animals) are considered adverse. At higher doses of erythropoietin, animals develop increases in red cell mass, chronic blood hyperviscosity, vascular stasis, thrombosis, peripheral resistance and hypertension. These findings are clearly fatal and adverse.

7.4 Communicating NOAEL in Toxicity Studies

After evaluating the tissues, the study pathologist provides pathology data tables (individual and summary) and a report containing clear and concise descriptions of test article-related gross and microscopic changes. In studies with recovery phases, the report also mentions whether the test article-related changes are reversible (complete or partial) or progressive once administration of the compound has stopped. There are generally two forms of pathology report in toxicity studies: an integrated final study report (with no separate pathology report) and a separate pathology report appended to the final study report. Regardless of the format, it is important that the pathologist and other contributing scientists provide an integrated assessment of important study findings (e.g. in-life findings, haematology, clinical chemistry, gross pathology (including organ weights), histopathology findings) in the executive summary and discussion of the study report. This will ensure that all test article-related pathology findings are accurately described and correlated with other study findings.

Whilst characterising the target organ toxicity, pathologists and study personnel should provide comments about the nature of the toxicity (adverse or nonadverse), which can provide justification for setting up a NOAEL at the study level. The NOAEL should not be discussed in the pathology report (Gosselin et al., 2011). In dose range-finding (short-term toleration) studies, it is not always necessary to determine the adversity of the pathology findings. The purpose of these studies is mainly to select doses for longer-term toxicology studies and assess the potential target tissues of the toxicity.

Further, a NOAEL should be determined for a toxicity study as a whole, rather than for individual pathology findings. Since a toxicity study is performed with a unique set of study conditions (e.g. duration, doses, route of administration etc.), the NOAEL is limited to these conditions and to the particular species under study. The NOAEL should be allowed to change as additional data are gathered for a particular compound. This allows for an all-inclusive monitoring of the temporal relationships of compound-related effects as data from longer-term studies become available.

7.5   Conclusion

There are many definitions of 'adversity' in nonclinical toxicology studies, and many perspectives on how to attribute 'adversity' to pathology findings. The recommendation is that determination of adversity in nonclinical studies should be based upon the outcomes given the conditions of the study, whether effects are harmful to the test species and interpretation of all compiled concurrent study data. Adversity determination should not be based upon extrapolation across species (including relevance to humans), intended human therapeutic indications or patient populations, presumption due to exaggerated pharmacology activity, assumption of primary, secondary or tertiary effects (including adaptive effects) or application of statistical analysis alone.

The adverse or nonadverse nature of test article-related pathology finding should be clearly stated and justified in the pathology report. The NOAEL should be determined at the study level and supported by the information in the study subreports, including the pathology report. Expert opinion and the judgement of subject-matter experts (including toxicologists, study personnel and pathologists) should be critical in assessing and communicating human risk assessment of adverse findings.

References

Bailey, S.A., Zidell, R.H. and Perry, R.W. (2004) Relationships between organ weight and body/brain weight in the rat: what is the best analytical endpoint? *Toxicologic Pathology*, 32, 448–66.

Derelanko, M.J. and Auletta, C.S. (2014) Handbook of Toxicology, 3rd edn, Taylor & Francis, Boca Raton, FL.

Dorato, M.A. and Engelhardt, J.A. (2005) The no-observed-adverse-effect-level in drug safety evaluations: use, issues, and definition(s). *Regulatory Toxicology and Pharmacology*, 42(3), 265–74.

ECETOC. (2002) Recognition of, and Differentiation between, *Adverse and Non-adverse Effects in Toxicology Studies. European Centre for Ecotoxicology and Toxicology of Chemicals. Technical Report No. 85.*

Eaton, D.L. and Gilbert, S.G. (2008) Principle of toxicology. In: Klassen, C.D. (ed.). Casaret and Doll's Toxicology: The Basic Science of Poisons, 7th edn, McGraw-Hill, New York, pp. 11–43.

EMA. (2004) Epoetin Delta Review. *European Public Assessment Report (Scientific Discussion).* Available from: http://www.ema.europa.eu/docs/en_GB/document_library/EPAR_-_ Scientific_Discussion/human/000372/WC500054474.pdf (last accessed July 29, 2016).

EPA. (2000) The Use of Data on Cholinesterase Inhibition for Risk Assessments of Organophosphorous and Carbamate Pesticides. US Environmental Protection Agency, Washington, DC. Available from: http://www.epa.gov/sites/production/files/2015-07/ documents/cholin.pdf (last accessed July 29, 2016).

EPA. (2007) Integrated Risk Information System (IRIS): Glossary of IRIS Terms. US Environmental Protection Agency, Washington, DC.

EPA. (2012) Sustainable Futures/P2 Framework Manual. EPA-748-B12-001, USEPA, Office of Chemical Safety and Pollution Prevention. US Environmental Protection Agency, Washington, DC.

Everds, N.E., Snyder, P.W., Bailey, K.L., Bolon, B., Creasy, D.M., Foley, G.L., Rosol, T.J. and Sellers, T. (2013) Interpreting stress responses during routine toxicity studies: A review of the biology, impact, and assessment. *Toxicologic Pathology*, 41, 560–614.

FDA. (2005) Guidance for Industry: Estimating the Maximum Safe Dose in Initial Clinical Trials for Therapeutics in Adult Healthy Volunteers. US Food and Drug Administration, Rockville, IN. Available from: http://www.fda.gov/downloads/Drugs/ GuidanceComplianceRegulatoryInformation/Guidances/ucm078932.pdf (last accessed July 29, 2016).

Filipsson, A.F., Sand, S., Nilsson, J. and Victorin, K. (2003) The benchmark dose method-review of available models, and recommendations for application in health risk assessment. *Critical Reviews in Toxicology*, 33(5), 505–54.

Gosselin, S.J., Palate, B., Parker, G.A., Engelhardt, J.A., Hardisty, J.F., McDorman, K.S., Tellier, P.A. and Silverman, L.R. (2011) Industry-contract research organization pathology interactions: a perspective of contract research organizations in producing the best quality pathology report. *Toxicologic Pathology*, 39, 422–8.

ICH. (2009) ICH Topic M3 (R2) – Non-clinical Safety Studies for the Conduct of Human Clinical Trials and Marketing Authorization for Pharmaceuticals. EMEA/CPMP/ ICH/286/95. Committee for Proprietary Medical Products.

IPCS. (2004) IPCS Harmoninzation Project: Risk Assessment Terminology. Part 1: IPCS/ OECD Key Generic Terms used in Chemical Hazard/Risk Assessment Part 2: IPCS Glossary of Key Exposure Assessment Terminology. World Health Organization.

Kaufmann, W., Bader, R., Ernst, H., Harada, T., Hardisty, J., Kittel, B., Kolling, A., Pino, M., Renne, R., Rittinghausen, S., Schulte, A., Wöhrmann, T. and Rosenbruch, M. (2009) 1st International ESTP Expert Workshop: 'larynx squamous metaplasia'. A re-consideration of morphology and diagnostic approaches in rodent studies and its relevance for human risk assessment. *Experimental and Toxicologic Pathology*, 61(6), 591–603.

Keller, D.A., Juberg, D.R., Catlin, N., Farland, W.H., Hess, F.G., Wolf, D.C. and Doerrer, N.G. (2012) Identification and characterization of adverse effects in 21st century toxicology. *Toxicological Sciences*, 126(2), 291–7.

Kerlin, R., Bolon, B., Burkhardt, J., Francke, S., Greaves, P., Meador, V. and Popp, J. (2015) Recommended ('Best') Practices for Determining, Communicating, and Using Adverse Effect Data from Nonclinical Studies. STP Draft Document 03/02/2015.

Klaassen, C. (ed.). (2013) Casarett & Doull's Toxicology, 8th edn, McGraw-Hill Education/Medical, New York.

Lewis, R.W., Billington, R., Debryune, E., Gamer, A., Lang, B. and Carpanini, F. (2002) Recognition of adverse and nonadverse effects in toxicity studies. *Toxicologic Pathology*, 30(1), 66–74.

Nikula, K.J., McCartney, J.E., McGovern, T., Miller, G.K., Odin, M., Pino, M.V. and Reed, M.D. (2014) STP position paper: interpreting the significance of increased alveolar macrophages in rodents following inhalation of pharmaceutical materials. *Toxicologic Pathology*, 42(3), 472–86.

NRC. (2007) Toxicity Testing in the Twenty-First Century: A Vision and a Strategy, National Research Council, National Academies Press, Washington, DC.

Parker, G.A. and Gibson, W.B. (1995) Liver lesions in rats associated with wrapping of the torso. *Toxicologic Pathology*, 23, 507–12.

Perry, R., Farris, G., Bienvenu, J.G., Dean, C. Jr, Foley, G., Mahrt, C. and Short, B. (2013) Society of Toxicologic Pathology position paper on best practices on recovery studies: the role of the anatomic pathologist. *Toxicologic Pathology*, 41, 1159–69.

Reasor, M.J. and Kacew, S. (2001) Drug-induced phospholipidosis: are there functional consequences? *Experimental Biology and Medicine*, 226, 825–30.

Reasor, M.J., McCloud, C.M., DiMatteo, M., Schafer, R., Ima, A. and Lemaire, I. (1996) Effects of amiodarone-induced phospholipidosis on pulmonary host defense functions in rats. *Proceedings of the Society for Experimental Biology and Medicine*, 211, 346–52.

Sergeant, A. (2002) Ecological risk assessment: history and fundamentals. In: Paustenbach, D. (ed.). Human and Ecological Risk Assessement: Theory and Practice, John Wiley & Sons, New York, pp. 369–442.

Shackelford, C., Long, G., Wolf, J., Okerberg, C. and Herbert, R. (2002) Qualitative and quantitative analysis of nonneoplastic lesions in toxicology studies. *Toxicologic Pathology*, 30, 93–6.

8

Limitations of Pathology and Animal Models
Natasha Neef

Vertex Pharmaceuticals, Boston, MA, USA

Learning Objectives

- Understand the limitations of animal models.
- Understand the limitations of disease models in testing compound efficacy.
- Understand the limitations of standard acute and chronic animal studies in determining safety.
- Understand the subjective nature of the data and potential for pathologist error.

A clear understanding of the limitations of pathology and of the animal models that produce pathology end points is a key skill for toxicologists and study personnel. In routine regulatory toxicology studies, the anatomic and/or clinical pathology end points are frequently the most significant elements of the study data, and both the data and the pathologist's interpretation of them can profoundly influence the subsequent development and use of the test article. Moreover, inappropriate generation and/or interpretation of pathology data can put humans at risk, waste animals, time and money and/or confound compound development. The purpose of this chapter is to provide an overview of the limitations of pathology and animal models, in order to allow toxicologists and study personnel to critically evaluate their pathology data and utilise pathology end points judiciously. The material presented reflects the personal opinions of the author.

8.1 Limitations of In Vivo Animal Models

8.1.1 Traditional Laboratory Species Used as General Toxicology Models

Routine general toxicology studies, as required by regulatory agencies, utilise young, healthy animals from outbred laboratory species. The design of these studies typically follows a standard format (Adams and Crabbs, 2013; Greaves et al., 2004), where the

Pathology for Toxicologists: Principles and Practices of Laboratory Animal Pathology for Study Personnel,
First Edition. Edited by Elizabeth McInnes.
© 2017 John Wiley & Sons Ltd. Published 2017 by John Wiley & Sons Ltd.

pathology portion uses a wide range of haematology, clinical chemistry and urinalysis end points and samples essentially all tissues for microscopic examination. Many of the limitations of these models are self-evident; the principal one is that animal physiology differs from that of humans, and therefore some toxicities observed in animal models may not be relevant for humans and some human toxicities may not be detected in animal studies (for a general review that includes summaries of concordance of human and animal toxicology data, see Greaves et al., 2004; Olson et al., 2000). Beyond this general principle, however, there a number of more or less obvious shortcomings of these models that toxicologists should bear in mind when assessing pathology findings (or absence thereof).

8.1.2 The Test Article May Not have Sufficient Pharmacological Activity in Routine Toxicology Species

In order for toxicology studies of substances such as pharmaceuticals (which act pharmacologically to achieve their desired effects) to correctly model human risk, the pharmacological activity of the substance in question at the doses used in the toxicology species must be at least broadly comparable with that in humans. This is most often a problem with biologic drugs, where monoclonal antibodies or other proteins can have very limited cross-species pharmacological activity. Similar target binding profiles between species are not sufficient to ensure this, since activity is sometimes also mediated by other parts of the molecule (such as the Fc portion of monoclonal antibodies), where binding and/or activity may differ between species, or the target itself may be distributed differently in the test species compared with humans. A good example of the limitations of animal models in this respect is the tragic outcome of a first-in-human clinical trial of a monoclonal antibody, TGN1412 – a humanised CD28 agonist antibody intended to act as a selective stimulant of regulatory T-cell expansion for the treatment of autoimmune diseases (Suntharalingam et al., 2006). In this case, despite comparable binding of TGN1412 to human and monkey CD28, TGN1412 did not demonstrate pharmacological activity in the cynomolgus monkey that was used as the primary toxicology species (Horvath et al., 2012). The monkey toxicology study thus failed to reproduce the cytokine storm that occurred in the human-trial subjects, leaving some with permanent disabilities. It later emerged that the reason for the lack of pharmacological activity in the cynomolgus monkey was that the main cell type responsible for eliciting the cytokine storm in humans (CD4+ effector memory T-cells) does not express the CD28 receptor in the cynomolgus monkey (Eastwood et al., 2010). Thus, even comparable target binding of the antibody in both humans and monkeys was insufficient to ensure pharmacological activity in the monkey, since the CD28 target was distributed on different T-cell subsets and performed a different function in humans.

Typically, biologic drugs with little or no activity in at least one toxicology species are very difficult to develop, and alternatives need to be found wherever possible. One approach is to create a surrogate test article that is active in a toxicology species. This is usually preferable to 'humanised' animal models (usually mice) that have been genetically modified to respond to the test article, since such models are usually poorly characterised and potentially still subject to other species differences, which may under- or even overestimate any adverse effects of the test article.

8.1.3 The Model May Not Identify Hazards Related to Causation or Exacerbation of Pathology that is Unique to Humans or Undetectable in Animals

There are a number of human diseases and other pathological conditions that are not meaningfully reproducible in standard toxicology species and for which routine studies are unlikely to be predictive. The most important example is exacerbation or precipitation of acute events related to human atherosclerotic cardiovascular disease. Nonsteroidal anti-inflammatory drugs (NSAIDs), particularly the relatively new selective COX2 inhibitors, are good examples of drugs that were extensively characterised in preclinical toxicology studies, but for which the cardiovascular disease liability for humans (always a theoretical possibility based on the mechanism of action of these drugs: Hawkey, 1999; Graham, 2006;) was widely recognised only post-approval using human data (Cannon and Cannon, 2012). Some of the precipitating causes of these human cardiovascular events were subsequently modelled in mice (Yu et al., 2012), but in the absence of the predisposing atherosclerotic vascular compromise, the human risk could not be evaluated directly using animal toxicology models.

Other significant examples include behavioural changes such as dysphoria and suicide ideation that are likely human-specific but in any case are not appreciable in laboratory species in the context of general toxicology studies. These types of finding in humans led to the withdrawal of rimonabant, the cannabinoid receptor-1 antagonist from the European market 2 years after its approval (Christensen et al., 2007).

The archetypal significant toxicity that is not generally predicted by preclinical toxicology studies is idiosyncratic drug-induced liver injury (DILI), which is a common cause of post-approval drug withdrawal, since it typically affects relatively few individuals and so is often not detectable in human clinical trials. It is similarly not generally observed in preclinical toxicology studies, despite the use of high doses in these studies (Greaves et al., 2004). Human-specific genetic factors likely play a role in individual susceptibility, and there is no widely accepted animal model (FDA, 2009; Daly and Day, 2012).

8.1.4 The Model May Not Identify Hazards with Low Incidence/Low Severity

Toxicology studies, particularly non-rodent studies, necessarily use limited numbers of animals – almost always fewer than the total number of humans potentially exposed to the test article in question. Whilst the use of high doses in these studies is intended to exaggerate toxicities in part to compensate for low animal numbers, it is still possible that even over multiple studies, toxicity issues can remain undetected or occur so sporadically that they are not attributed to the test article. Where the 'true' incidence of a particular finding is less than about 10–20% in rodents or 30–50% in large animals, it is perfectly possible that, by chance, a finding to which the toxicology species is susceptible would not occur at the highest dose in a standard study, and that if it occurred at a lower dose, it might be dismissed as not test article-related based on the lack of a dose response. This is particularly likely in non-rodent studies using small numbers of animals, and a common mistake in interpreting such data is to assume that the absence of a finding that occurred at a low incidence in rats (despite similar systemic exposures) constitutes a 'rodent-specific' finding. In these cases, it is wise to bear in mind that that with low animal numbers, absence of evidence is not necessarily evidence of absence.

8.1.5 Potential for Misinterpretation of Reversibility/Recovery for Low-Incidence Findings

A related problem in toxicology studies is overinterpretation of recovery animal data, particularly for non-rodents, which typically utilise group sizes of just two animals/sex/ dose for the recovery arm. This can also occur with low incidence findings in rodents, where only five animals/sex is typical for recovery groups. The absence from recovery animals of a pathology finding that occurred in only a minority of animals sacrificed at the end of the dosing period may be simply because it was not present in any of the recovery animals in the first place, but toxicology reports sometimes interpret such data as unequivocally indicating 'reversibility'. Reversibility from clinical pathology findings in both rodents and non-rodents is also sometimes claimed in toxicology studies, even when a glance at the data shows that the individual animals used for recovery were not affected at the end of the dosing period. In the case of anatomic pathology lesions, the presence of a clinical pathology biomarker for the lesion in question is useful in supporting an interpretation of reversibility, as is an understanding of the nature of the lesion and its likely reversibility (Perry et al., 2013).

8.1.6 Potential for Over- or Underestimation of the Relationship to Test Article of Findings that have High Spontaneous Incidence in Laboratory Species, but are Relatively Rare in Humans

The toxicologist should be aware of common spontaneous background changes in the different laboratory species, since chance variation in their incidence amongst the dose groups can mimic or mask a test-article effect. A common situation is the misdesignation of a finding as a test article-related when there is an increase in severity or incidence over that in the control group that is actually just due to chance, or when a finding is known to be a common background change but by chance does not appear at all in the control group in that particular study. These are difficult judgements to make, because genuine test article-related increases in findings that are indistinguishable from spontaneous findings can occur. However, an awareness by the toxicologist of the most common types of spontaneous pathology lesions (Chapter 3) can alert him or her to cases where further investigation – such as review of historical control data (to understand what constitutes a normal incidence of the finding in question) or obtaining a third-party expert opinion – may be appropriate to ensure the best possible interpretation of the data.

Reviews of spontaneous pathology in all the common laboratory-animal species are published fairly frequently (e.g. Lowenstine, 2003; Chamanza et al., 2010; McInnes, 2012). Typical examples include cardiomyopathy in rats and focal renal tubular degeneration/regeneration in most species; there is plenty of scope in these cases for random differences in incidence/severity between groups to either mask or falsely suggest a potentially serious test-article effect. This can be a particular problem in non-rodent studies where the number of animals is small.

A useful comparison for these situations is carcinogenicity studies, in which statistical analysis with correction for multiple testing and comparison with historical control data is performed routinely to prevent overinterpretation of differences in tumour incidence between treated and control animals (FDA, 2001). This means that, particularly for common lesions, p-values well below the traditional 0.05 cutoff are required for meaningful statistical significance. Since statistical analysis is not usually performed for non-neoplastic lesions in general toxicology studies, adjustment of this kind does not take place automatically, and overinterpretation of data is thus more likely.

Another limitation to consider is that comparison of incidences of putative test article-related findings with historical control data for general toxicology studies is sometimes unreliable, since the very fact that these findings are non-neoplastic and can be 'normal' means that many pathologists do not diagnose them routinely, especially in control animals, where they cannot be test article-related (McInnes and Scudamore, 2014). In these circumstances, historical control data can give a falsely low impression of the true background incidence of the finding. For toxicologists dealing with potential overinterpretation of pathology findings that are not explainable using historical control data, the best course (if this has not been done already) is to request comparison of the sections in question with a reasonable number of control animals from other recent studies, with review by both the study pathologist and the peer reviewer. If this is not possible, review of the slides by a third-party pathologist with a very large amount of experience reading studies of that particular duration in that species, may be justified to ensure the most appropriate interpretation of the data.

8.1.7 Exclusive Use of Young, Healthy Animals Kept in Ideal Conditions Gives Limited Predictivity for Aged/Diseased Human Populations

Minimising interanimal variability in order to improve the sensitivity of preclinical toxicology studies is one of the main reasons for using adolescent or young-adult animals that are free of intercurrent disease or ageing changes, and for standardising their environment, diet and other experimental conditions. These healthy animals generally have a large reserve functional capacity within major organ systems (immune, cardiovascular, renal, hepatic etc.), and so in many situations will be less likely to manifest evidence of organ malfunction in the presence of degenerative and/or functional changes mediated by a test article. In contrast, the general human population outside a clinical trial setting will contain many individuals whose susceptibility would be significantly greater than that of the test animal population, and an intended patient population will often contain a disproportionate number of individuals compromised in one or more respects. Again, the higher doses used in toxicology studies relative to anticipated human exposure will compensate to some extent for the high reserve functional capacity of the animal subjects, but in some situations – such as when overall tolerability issues prevent use of good exposure multiples in toxicology species – the use of healthy animals kept in ideal conditions is potentially a significant limitation of general toxicity studies.

Conversely, some toxicities can be exaggerated in young, growing animals, leading to an overestimation of the risk in an adult human patient population. A common example is compounds that affect bone deposition or remodelling, which can produce quite profound changes in young growing animals with active bone growth plates but have little or no effect in older human adults (Gunson et al., 2013). Typically, the more rapidly growing the test animal, the greater its sensitivity, which can lead to the seemingly counterintuitive situation where toxicities are more prominent in shorter-term studies than in longer-term ones conducted at the same doses, simply because at study termination, the animals are younger and more rapidly growing in the shorter-term studies.

Another limitation imposed by the use of healthy animals is that the pharmacology of the test article may become dose-limiting when it is intended to counteract abnormal physiology in the intended patient population. Examples include hypoglycaemic and hypotensive therapies that cause life-threatening hypotension/hypoglycaemia at low systemic exposure multiples in normotensive or normoglycaemic animals. This precludes observation of any other toxic features of the molecules that could occur in patients, who can tolerate higher doses.

Finally, a widely recognised limitation that can be difficult for toxicologists, clinicians and regulators to evaluate based on the data they receive in the study reports is the issue of the relative sexual immaturity of the young males in non-rodent studies confounding assessment of male reproductive-organ toxicity. Immature testes that are in a quiescent, prepubertal state will in most cases be less susceptible to reproductive toxicants, whilst peripubertal testes and epididymes frequently demonstrate evidence of aborted spermatogenesis and frank degeneration that are indistinguishable from potential test article-related toxicities (Creasy, 2003). Testicular changes related to peripubertal status may thus mask or falsely suggest test-article effects. Unfortunately, it is common for pathologists not to record immaturity or peripubertal status in these non-rodent studies, and thus the toxicologist or regulator may be none the wiser as to whether a study has adequately evaluated testicular safety or might be confounded by peripubertal changes within the testes. The age of the individual animals in the study is a useful guide to likely sexual maturity (minimum 5 years in the cynomolgus monkey (Smedley et al., 2002) and 10–12 months in the dog (Lanning et al., 2002; Creasy, 2003)), but individual animal ages are usually not documented in study reports. A simpler and more reliable way for the toxicologist to estimate sexual maturity in a toxicology study is to use testicular weight (as a surrogate for volume), since individual animal testes weights are normally readily available in the study report. In the cynomolgus monkey, a combined testicular weight of approximately ≥20 g suggests sexual maturity (Ku et al., 2010); values in the dog are more variable, but testes weighing >20 g are certainly likely to be mature (Olar et al., 1983; Goedken et al., 2008).

8.2 Efficacy/Disease Models as Toxicology Models

Use of in vivo animal-disease models (that are traditionally used to evaluate the efficacy of candidate therapies as toxicology models) is becoming more widespread in the pharmaceutical industry and, under certain special circumstances, is being considered actively by regulators as a means of evaluating potential drug toxicities encountered in the clinic that would not be identified in traditional preclinical toxicology studies (FDA, 2014). The range of potential models is very diverse, and includes genetically modified rodents, surgical models such as ureter ligation to produce renal insufficiency (Chevalier et al., 2009), meniscectomy to produce arthritis of the knee (Bendele and White, 1987), and disease states produced by known toxicants such as streptozotocin as a model for diabetes mellitus (Like and Rossini, 1976), or adjuvant-induced arthritis (Bendele et al, 1999). Such disease models have been used successfully to investigate toxicities emerging in the clinic: for example, a genetically modified mouse model of Alzheimer's disease (APP23 transgenic mice) was used to investigate the pathogenesis of fatal meningoencephalitis occurring in humans with deposition of β-amyloid in the cerebral vasculature, which had been immunised with an amyloid-β in a clinical trial (Pfeifer et al., 2002).

The potential advantages of collecting safety data from animal-disease models include minimising the use of animals by collecting both safety and efficacy endpoints from the same set of study animals and obtaining toxicology information on particular therapeutic targets/candidate molecules at a very early stage in development. Mice that are genetically modified to lack the pharmacological target of the drug can also sometimes help to distinguish toxicities related to the intended pharmacology of the test article. Other possibilities include using supplementary toxicology models to overcome the problem of exaggerated pharmacology confounding the interpretation of data collected in routine toxicology studies. Examples include hypoglycaemic therapies such as synthetic insulins whose toxicology cannot be explored in normal animals due to life-threatening hypoglycaemia at low systemic exposure multiples (FDA, 2000). Even where reasonable drug exposures can be achieved in normal animals, pathologies due to exaggerated pharmacology can emerge, and special animal models are required to demonstrate a lack of such findings where the model resembles the target patient population. One example is glucokinase-agonist drugs, which were intended as hypoglycaemic agents for the treatment of diabetes mellitus; these produce vascular and neurological pathology findings in monkeys and rats, respectively, related to hypoglycemia that would occur rarely or not at all in the intended patient population (Pettersen et al., 2014; Tirmenstein et al., 2015).

That said, because of the limitations of these disease models (described below) they are not generally used as toxicology models, even where their use might permit higher exposures than would be tolerated by normal animals. Instead, they are more often used in directed studies to provide experimental evidence where this is needed to support a hypothesis that a particular toxicity will not be relevant for the intended use of the test article in humans (Morgan et al., 2013).

Animal models of disease could also theoretically also be used to determine whether the toxic effects of a test article might be accentuated in special patient populations. They might be used for drugs producing minor perturbations in organ function that have few or no adverse effects in standard toxicology studies but which might represent a significant risk for patients with pre-existing compromise of the organ system in question (most commonly immune, cardiovascular or renal). In practice, however, disease models are seldom used for this type of safety testing, and it is more common to evaluate test-article effects on immune, cardiovascular or renal function of potential relevance for special patient populations using directed specialised studies in healthy animals, since (for reasons outlined below) these are likely to be more sensitive and to give more reproducible results. The well-established in vivo safety pharmacology test systems in normal animals that are routinely used for the sensitive detection of functional changes in renal, gastrointestinal, respiratory or cardiovascular systems following single doses can be adapted, if necessary, for multiple doses, and this approach is suggested by regulators for the assessment of risk in special patient populations (ICH, 2001). Similarly, the immunotoxicity testing protocols recommended by regulators generally use healthy animals for the detection of small changes of potential relevance to immunosuppressed patient populations (ICH, 2006), rather than looking for changes in an animal model of immunosuppression.

8.3   Limitations of Efficacy/Disease Models as Toxicology Models

As already mentioned, there are many compelling theoretical reasons (scientific and nonscientific) why animal-disease models might be useful for toxicology data collection. Various examples of the successful elucidation of toxicity risk using these models are published in the literature (as reviewed from a toxicologic-pathology perspective by Morgan et al., 2013). However, published work typically does not reflect situations in which the hypothesis was not confirmed in the animal disease model or where other potential safety signals of uncertain relationship to the test article or uncertain relevance for humans appeared in these models. Thus, in practice, there are a variety of issues that need to be carefully considered by the toxicologist before embarking on the collection of safety end points from such studies. Neglecting to do so may result in generation of data that are misleading or uninterpretable but nonetheless must be submitted to regulatory authorities.

8.3.1   Lack of Validation as Safety/Toxicology Models

If they are validated at all, animal models of disease are usually validated as models for showing the efficacy of therapies intended to improve the disease state. This is very different from showing that a test article does not make the disease worse, or from looking for toxic effects in organs that are not directly affected by the disease (but which may be indirectly affected by the disease state, which complicates interpretation of any pathology). At the very least, some ad hoc validation with positive and negative controls (compounds currently used in the human disease being modelled that do or do not, respectively, have particular toxic liabilities in that particular population) is advisable. Without this, such a model will not provide a credible assurance of safety, nor should any identified adverse effects necessarily be considered translatable to a human population. Even then, validation in any universally meaningful sense would require the use of many positive and negative controls, utilising different pharmacological mechanisms of action, which even if available, would take many years and require prohibitive numbers of animals.

A good example of the difficulties of validating a disease model for the determination of toxic effects related to enhancement of a disease is the well-established 4-hydroxybutyl(butyl)nitrosamine model of bladder cancer in rodents, which was recently recommended by the US Food and Drug Administration (FDA) for the assessment of the potential for a pharmaceutical to promote bladder cancer in humans with existing preneoplastic bladder pathology (FDA, 2013). Various agents have been identified that both promote and suppress bladder carcinogenesis in this model, including some genotoxic agents that are believed to play similar roles in human bladder carcinogenesis (Wanibuchi et al., 1996). Unfortunately, agents that have not been associated with human bladder cancer (despite extensive monitoring of exposed populations, in some cases) have been found to promote bladder cancer in this model. These include ascorbic acid (vitamin C) (Fukushima et al., 1984) and the antidiabetic drug, rosiglitazone (Lubet et al., 2008). Furthermore, agents known to increase the risk of human bladder cancer, such as cyclophosphamide, have not acted as promoters in this model (Babaya et al., 1987). Hence, despite this being a well-established and well-characterised cancer model, its utility as a predictor of human safety with respect to the promotion of bladder cancer by nongenotoxic agents is doubtful. It is also useful to note that the collection of the data needed to reach this conclusion has taken many years and large numbers of animals; this kind of work would be very difficult to undertake de novo in order to validate another animal model of disease as a reliable predictor of human safety.

In practice, the difficulties inherent in fully understanding animal models of disease and how safety data obtained from them may translate to humans mean that once baseline safety data have been obtained in routine toxicology models, further safety data pertaining to specialised disease situations or special human patient populations are more reliably obtained from human trials.

8.3.2 Disease Models Rarely Have All the Elements of the Equivalent Human Disease

It is well accepted that animal models of disease seldom have all the elements of the equivalent human disease. This limits their applicability to safety screening – in many respects, much more than it does for efficacy screening, where a single predefined and validated end point can be selected, regardless of the lack of translatability of the other features of the disease in that particular model.

Type 2 diabetes is a good example in which there are a variety of models that have at least some features of the human disease, but none has all the features (including, of course, the accelerated atherosclerosis noted in human diabetes populations), and diabetes therapies that are effective in humans do not necessarily show efficacy in every model (reviewed by Calcutt et al., 2009; King, 2012). For this reason, none of these models would be likely to provide a reliable indication of all potential toxicities in the human diabetic population, and may demonstrate irrelevant toxicities that are unique to that particular disease model in that particular species.

Of course, type 2 diabetes is a complex disease and its underlying aetiology is not well understood, so the shortcomings of the various animal models are perhaps not surprising. However, even diseases with a very well understood pathogenesis are frequently not well modelled in animals. Cystic fibrosis is a monogenic disease in which a variety of mutations in the CFTR chloride channel gene produce a very consistent pattern of pathology in humans. However, the mice, pigs and ferrets with natural or engineered mutations in the corresponding CFTR channel that can be used as in vivo models do not fully recapitulate the disease as it is manifest in humans (Keiser and Engelhardt, 2011).

8.3.3 Limited Sensitivity Produced by Increased Interanimal Variability amongst Diseased Animals and/or Low Animal Numbers

Typically, the pathology produced in animal models of disease will be quite severe, frequently progressing over durations relevant for toxicity studies, with a fair amount of interanimal variability (almost certainly more at baseline than would be expected in a similarly sized group of healthy animals in a standard toxicology study). This is particularly true for surgical models of disease. Since efficacy effects need to be readily demonstrable in most or all animals at doses/exposures comparable to those to be used in humans, lower sensitivity introduced by greater variability is acceptable in efficacy models. However, a relatively small effect occurring in one or a few animals may be significant in a toxicology study; such an effect may be lost in the background variability of an animal disease model or, alternatively, chance findings related to the model and not the test article may be interpreted as test article-related.

An additional complication is that progression of disease in the model over the course of a study needs to be considered. This will add to variability even if the animals were originally randomised based on a disease metric. Disease progression (and hence other toxicity end points) may be influenced by the test article, and (of course) recovery from toxicity end points in a clinically deteriorating animal will be very difficult to assess. It should also be noted that the disease process and/or the strain of animal (often inbred) may affect test-article pharmacokinetics and metabolism such that toxicokinetic data need to be evaluated before disease-model data can be compared with data obtained from traditional toxicology studies. The usefulness of pooled samples for TK analysis from a group of individual animals with disease of differing severity also needs to be considered.

Such inherent additional variability in disease models indicates that the animal numbers needed to reliably assess toxicity end points should theoretically be higher – potentially, much higher – than the numbers used in routine toxicology models if the studies are to be similarly powered. In practice, however, the logistical difficulties inherent in models such as surgery or the generation of transgenic animals often means that such numbers cannot be achieved, and the studies are correspondingly less sensitive.

8.3.4 Lack of Historical Data

Historical data on safety end points are usually limited or even absent from disease models. This means that chance findings may be overinterpreted by the pathologist or the regulators as test article-related, especially in studies with small numbers and high variability amongst animals.

8.3.5 Risk Associated with Nonregulated Laboratory Conditions

In contrast to traditional toxicology studies, studies involving surgical models and the use of nonstandard animal species/strains must often take place in facilities that have no experience complying with Good Laboratory Practice (GLP) and may have limited disease control, procedure standardisation, data recording, equipment calibration, staff qualifications and so on. Any one of these factors may affect the quality and integrity of data. In these circumstances, the risk of spurious findings (including death), or of significant findings being undetected, is higher. For safety end points, even a single abnormal finding must be considered potentially test article-related if there is no convincing alternative explanation for it, and must be reported to regulators if the compound is in the clinic or is subsequently developed. Toxicologists should therefore carefully consider the risks of generating data in these nonregulated environments.

With respect to pathology in these settings, it is absolutely critical to ensure that any pathology end points are interpreted by a qualified and experienced anatomic and/or clinical pathologist, as appropriate. This should apply even when pathology data for the model in question are usually generated by nonpathologists. There are many examples of histopathology misinterpretation by nonpathologists in the scientific literature, even of familiar, predefined efficacy end points (as reviewed by Ince et al., 2008), and the likelihood that unfamiliar end points will be reliably identified and/or interpreted by an untrained pathologist in unfamiliar tissues is slim.

Even qualified pathologists in academic institutions may not be familiar with toxicology data-collection and naming conventions. Issues that can arise include replacing primary observations with diagnoses or interpretations: what should be described as yellow discolouration of the tissues becomes jaundice, for example. These interpretations within a necropsy report unfortunately become raw data and cannot be readily reinterpreted, even if other data from the study justify it (e.g. the fact that the test article imparted a yellow colour to the tissues). Pathologists working primarily on specific projects in an academic setting may also be relatively unfamiliar with some of the tissues examined as part of the comprehensive tissue list required for toxicology screening, and hence may misinterpret or fail to recognize findings of toxicological significance.

In summary, whilst animal models of disease can be successfully used in the development of new therapies, they are only rarely suitable for generating toxicology data. They are most appropriate for answering specific questions with predefined end points, rather than general toxicology evaluation – if that objective cannot be fulfilled using conventional toxicology or safety pharmacology models. Demonstrating efficacy requires an effect large enough, in a sufficient number of animals, that the changes are statistically significant. The desired end point(s) and the baseline variability in those end point(s) are known in advance, and the study can be powered appropriately. Toxicity, on the other hand, can be significant even when it affects a single animal, regardless of statistical significance, and there are an unlimited number of potential end points, which are not defined in advance. The inherent variability of disease models and the lack of historical data on these potential toxicity findings mean that distinguishing real test article-related toxicity findings from chance effects cannot be done reliably in most circumstances.

8.4 Limitations of Pathology within In Vivo Toxicology Models

8.4.1 Anatomic Pathology Evaluation Will Not Identify Hazards with No Morphological Correlates

This is self-evident, but it is useful for a toxicologist to appreciate the range of possible underlying toxicities when anatomic pathology examination of the tissues does not provide a primary cause of death/moribundity or other adverse clinical signs in a toxicology study. These include cardiac arrhythmias (for which an in vivo telemetered cardiovascular safety pharmacology study in the appropriate species will be helpful), certain neurological abnormalities (for which specifically designed safety pharmacology end points, nerve-conduction studies or even careful clinical observation can be more sensitive than anatomic pathology; Greaves, 2012) and changes in organ function at sites such as the renal tubular epithelium (for which urinalysis end points may be more informative). That said, pathologists are both clinically qualified and specifically trained in understanding the processes that lead to disease manifestations of all kinds, so a discussion with the study pathologist can often be helpful for toxicologists seeking mechanistic explanations for nonpathology study findings.

8.4.2   Limitations of Pathology when Evaluating Moribund Animals or Animals Found Dead on Study

Where a test article produces toxicities that result in death or moribundity, determining the primary toxicity that led to a chain of events producing other, secondary anatomic or clinical pathology changes can be particularly challenging to unravel from the collection of abnormalities presented to the pathologist. Poor organ perfusion related to terminal cardiac insufficiency or shock from other causes can result in necrotic changes in a variety of organs, as well as sharp increases in circulating hepatic and renal damage markers, independent of the primary target organs of toxicity (Kumar et al., 2015). Toxicologists should appreciate that the collection of anatomic and, particularly, clinical pathology data from moribund animals (e.g. at a non tolerated dose in a dose range finding study) may be misleading, especially if the data are not interpreted by the appropriate experts, or if communication is poor and the pathologist interpreting the data is unaware of the clinical status of the animals.

8.4.3   Limitations of Anatomic and/or Clinical Pathology End Points within other Types of In Vivo Preclinical Safety Study

Anatomic pathology end points are sometimes added to other toxicology or in vivo safety pharmacology studies (routine or investigative) in a desire for thoroughness when investigating a particular toxicity, or to reassure the toxicologist that a functional change has no anatomic pathology correlates. Other situations include the elucidation of unexpected mortality or other toxicity findings that emerge during the study. The limitations of the data obtained from such studies that should be carefully considered by the toxicologist include:

- Animals may be non-naïve, and pathology changes unrelated to treatment with the current test article may be misinterpreted.
- Appropriate control tissue may not be available for comparison, leading to false conclusions regarding the relationship of any changes noted to the test article.
- Tissue changes related to surgery, chronic instrumentation and so on may confound interpretation (especially without control animal data).
- Necropsies may be performed by members of staff who are unfamiliar with standardised necropsy protocols and procedures, who may take unsuitable samples or misinterpret or overlook test article-related, macroscopic findings.

Similarly, reproductive toxicology protocols sometimes provide for the recording of macroscopic pathology findings at necropsy, with or without subsequent microscopic evaluation of those tissues or clinical pathology end points. This may be useful on an ad hoc basis in a study in which there is unexpected moribundity or mortality, but data interpretation can be confounded by the presence of unfamiliar morphological changes related to the hormonal status of pregnancy or lactation (especially if corresponding control tissues are not available) and by a lack of historical anatomic pathology data or clinical pathology reference ranges for the applicable stage of pregnancy/lactation.

In situations where there is a need to use anatomic or clinical pathology in order to understand events occurring in studies in which these are no routine end points, a directed study to collect anatomic and clinical pathology data is often better conducted separately.

8.4.4 Limitations of Histopathology Related to Sampling Error

By its nature, histopathological examination evaluates a two-dimensional (2D) sample of an organ, which, depending on the size of the organ, can produce a section covering from 100% of its cross-sectional area (e.g. rodent thyroids, adrenals, pituitary) to much less than 1% (e.g. dog liver or lung, skeletal muscle in any species). As a percentage of the three-dimensional (3D) volume of a tissue or organ, of course, the area sampled is substantially less than this. Since many pathology lesions are focal in nature, their extent may be underestimated, they may be missed altogether or random variations in their distribution in these very limited samples can mean that true differences in extent or severity between groups are exaggerated or overlooked. Whilst it is generally appreciated that some potential for sampling error is inevitable, toxicologists (and regulators) are sometimes unaware of the extent to which pathology laboratory practice (no matter how rigorously compliant with GLP regulations) can contribute to sampling error, over- or underestimation of toxicity and/or inconsistency between studies.

Competency and experience of necropsy staff is a critical factor in ensuring recognition and sampling of focal gross lesions for subsequent microscopic examination and diagnosis. Necropsy supervision by a pathologist is helpful, but is not a substitute for experienced necropsy personnel: there is only so much a supervising pathologist can observe of a mouse carcass over the shoulder of the prosector, or when multiple animals are being necropsied simultaneously. Furthermore, toxicologic pathologists do not typically conduct necropsies themselves, and recently qualified pathologists may have relatively little experience with the gross appearance of organs from toxicology species. If a lesion is not recognised or is not sampled, this can result in significant over- or underestimation of human risk, depending on whether the lesion occurred in a control or a treated animal. Furthermore, necropsies are not conducted blind: prosectors are aware of which animals have been treated, and it is important that control animals are assessed with equal scrutiny to prevent any bias in the diagnosis and sampling of gross changes in treated versus control animals.

Another point to bear in mind is that many histology lesions reported in toxicology studies are not grossly visible and that lesion distribution may not be uniform throughout an organ. Thus, the sites within the tissues/organs from which histology samples are obtained need to be consistent and therefore specified in the laboratory's necropsy standard operating procedure (SOP). The level of detail within the necropsy SOP is a useful guide to the likely consistency of the end product of any particular laboratory. Certain liver toxicities are more or less apparent in particular lobes of the liver, for example (Richardson et al., 1986), and the site of sampling of skeletal muscle in rodents affects the relative numbers of type I and II fibres obtained in the section. Since different fibre types can respond differently to a given pharmacological agent, differences in muscle-sampling protocols can affect the manifestation of muscle toxicity (or lack thereof) in a toxicology study (e.g. Westwood et al., 2005; Okada et al., 2009).

Laboratory tissue-processing procedures can also significantly limit the value of the pathology data obtained. Even where there is meticulous consistency within any particular laboratory, interlaboratory differences in the amount of tissue present on the slide, its site of origin within the sampled organ and its orientation on the slide can affect the sensitivity of a study in detecting certain test article-related changes. For example, in this author's experience, testicular sections can vary between a full cross-section of each testis (split into two pieces if the testis is large, with all parts of both epididymes (head, body and tail)) and a segment approximately one-sixth of the cross-sectional area of each testis placed together on the same slide as the epididymal head and the contralateral epididymal tail. The latter, at perhaps 20–30% of the total tissue area compared with the former, is quicker for the pathologist to evaluate, but with a sensitivity for focal lesions obviously much less than where all parts of the testis and epididymis are examined microscopically. Thyroid tumour incidence, too, can be affected by whether the organ is sectioned longitudinally or transversely (Hardisty, 1985), and since there is no standard method, the data for certain thyroid changes may not be comparable between laboratories. Unfortunately, for the toxicologist, finding out the details of these sampling procedures will often entail tracking down the SOP in force at that particular laboratory at the time the study was conducted.

Other interstudy/interlaboratory differences include the consistency with which certain tissues are sampled at any level. It is common for the laboratory not to recut or resample wet tissue when small, easily missed organs are not present at the first attempt, including mammary gland in male rats, parathyroids in rats of both sexes, thymus gland in older rats and suborgans such as adrenal medulla and all three of the functionally separate segments of the pituitary gland in rats and mice. Further, some laboratories/pathologists may not record if these tissues or subtissues are not present in the section examined. Sampling differences of this type can potentially confound the detection of test article-related changes or (more often) result in different NOEL outcomes for the same study design between different laboratories. Where unexpected differences between studies emerge, toxicologists should review sampling protocols for tissues of concern in order to evaluate the possibility that tissue-sampling issues are confounding the anatomic pathology data.

8.4.5 Limitations of Quantitative Anatomic Pathology

It has become relatively simple and cheap to use image analysis to quantitate tissue changes on 2D microscopic sections, either directly from standard haematoxylin and eosin (H&E) sections or using immunohistochemistry, in situ hybridisation (ISH) or histochemical special stains. The data appear less subjective and more sensitive than those acquired from routine semiquantitative methods, which can be reassuring for toxicologists and regulators alike. Common quantitative pathology end points assessed in one or a few 2D microscopic sections include proliferation indices (using markers such as KI67 or BrdU), percentage cross-sectional area staining for a particular cell type (e.g. insulin-positive β-cells of the pancreatic islets as a percentage area of the entire section) and relative numbers of CD4+ versus CD20+ cells in a section of lymphoid tissue. Unfortunately, the validity of these end points depends on a number of assumptions, some or all of which are likely to be invalid when the data are acquired from 2D microscopic sections prepared using fixed material from routine toxicology studies (Mendis-Handagama and Ewing, 1990; Boyce et al., 2010a; Gundersen et al., 2013). The issue is particularly problematic where estimates of cell/nuclei number or density are involved, and pathologists themselves are sometimes not aware of some of the limitations of these techniques. An excellent review of some of the pitfalls is provided is provided by Boyce et al. (2010a) and Gundersen et al. (2013), but some of the key false assumptions that can produce misleading results are summarised here:

- The assumption that the profile counts of nuclei or cells being counted in a 2D section reflect their relative numbers in the 3D space of the organ. In fact, the counts in a 2D section are proportional to cell height perpendicular to the plane of section (which is unknown), and thus there is no known relationship to the actual counts within the organ as a whole.

- The related assumption that the test article does not affect the shape or size of the cell type or nucleus being counted, such that its appearance in profile in 2D sections is not biased in treated versus control animals. This is a particular problem in calculations involving labelling indices for cell proliferation where changes in rates of cell death or cell division are highly likely to result in changes in nuclear or cell size, and hence counts visible in a 2D section. In these cases, where effects on cell proliferation are small, the true test-article effect may be indistinguishable from effects on nuclear or cell height. The latter change will affect the likelihood of a particular nuclear profile appearing in the section: larger (and hence taller) nuclei will be oversampled, whilst smaller, shorter profiles will be undersampled.
- The assumption that in addition to changing the parameter of interest, the test article does not also affect other cell types present in the section that are being used as a reference when evaluating that parameter – the 'reference trap'. For instance, percentage insulin staining area of pancreas sections to determine whether a test article affects β-cell 'mass'. In a toxicology study, it is entirely possible that the volume of exocrine pancreatic tissue is diminished compared with controls – due either simply to lower body weight in treated animals or to a selective loss of exocrine tissue due to inappetence and reduced secretory demand, reduced blood flow and so on. If the β-cell area (insulin staining) is the same in both control and treated groups (say, 3 units – consistent with β-cells making up 1–5% of total pancreas tissue) and the total cross-sectional area of the tissue is 100 in the control group and 75 in the treated group (the latter due to reduced exocrine tissue as a direct or indirect effect of the test article) then the percentage cross-sectional area of pancreas staining with β-cells is 3% in controls and 4% in treated animals – an apparent 33% increase in β-cell 'mass', despite β-cell area not changing. Of course, multiplying by total pancreas weight will adjust for this false assumption (although even this basic adjustment is often not performed), but not for other potentially false assumptions, such as tissue shrinkage during processing, non-uniform distribution of islets or non-uniform distribution of test article-related effects in either the endocrine or the exocrine pancreas.
- The assumption that processing-related tissue shrinkage is not affected by the test article. It is possible for tissue shrinkage to occur disproportionately in treated animals, such that fixed organ weights or the tissue area on a slide will provide a biased estimate of tissue volume. This effect can be very marked: in a study of the effects of endocrine agents on testicular Leydig cell numbers (see later in this section), shrinkage of testis tissue from fresh to fixed state was approximately 1.4% in control rats versus 14% in treated rats, whilst further shrinkage following processing to slide was approximately 3% in control rats versus 17% in treated rats; thus, comparison between treated and control animals of fixed organ weights or cross-sectional areas on microscopic sections was invalid (Mendis-Handagama and Ewing, 1990).
- The assumption that a single section (usually taken through the thickest part of the organ) is representative of the organ as a whole, and the related assumption that the test article affects all areas of the organ uniformly.

The following is a simplified presentation of a published example of the pitfalls of attempting morphometry on 2D microscopic sections based on false assumptions. In this case, the effect of treatment with 17β estradiol on testicular Leydig cell numbers in the rat was examined. When Leydig cells were evaluated based on their numerical density (i.e. number of cells per unit area of testis, derived from a standard histological section through the testis), the test article appeared to dramatically *increase* Leydig cell number. In contrast, when appropriate stereological techniques were used and absolute number was evaluated, a dramatic decrease in Leydig cell numbers (as expected) was observed. The reasons for the discrepancy were as follows: the test article also decreased the volume of the testes (and hence cross-sectional area), which was used as the denominator for comparing cell density (the 'reference trap'); the processing shrinkage was greater in treated animals, and hence further magnified the artefactual increase in cell numbers based on cross-sectional area; and the test article induced shrinkage of the Leydig cells themselves, so that relative to the remaining tissue, their numbers as counted by profiles in 2D sections appeared increased (Mendis-Handagama and Ewing, 1990).

Generally speaking, if group differences are not readily appreciable by eye (obviating the need for morphometry), simple morphometry with image analysis of one or a few representative sections, without systematic stereological analysis, can give misleading results that do not reflect the actual changes occurring in a particular organ. Design-based stereology is normally the only appropriate option where quantitation of toxicologic anatomic pathology changes is required for decision-making (Boyce et al., 2010a). Adding simple morphometry to confirm qualitative or semiquantitative observations as a 'nice-to-have' may produce misleading results that contradict other data in the study and can result in false rejection of a valid hypothesis. A suggested rule of thumb (Boyce et al., 2010b) is that large changes (three- to fourfold) in common end points (e.g. comparative numerical density of Ki67-positive nuclear profiles to demonstrate hyperplasia) are unlikely to be masked by false assumptions, especially when supported by other data, such as a corresponding organ-weight increase. Smaller changes, however, are more likely to be influenced by biases, such that the real change may be directionally opposite from what is calculated using simple 2D morphometric data. That said, design-based stereology is relatively cumbersome and expensive, and is probably best used when the quantitation of subtle tissue changes is the only means of testing a hypothesis.

Toxicologists considering quantitative anatomic-pathology end points should question (i) is quantitation really necessary over and above standard semi-quantitative anatomic pathology evaluation? And, if quantitation is considered necessary, (ii) is analysis of microscopic sections the most appropriate method? It is sometimes tempting for pathologists to recommend morphometry as a quantitative extension of their original semiquantitative morphological diagnosis. However, the same objective can sometimes be achieved more directly, cheaply, quickly and effectively using alternative methods.

For example, accumulation of certain materials within cells or organs may be better evaluated directly by tissue assay than by staining them and attempting to quantitate that staining in sample 2D sections. For instance, fatty change within the liver can be quantitated by directly measuring total triglyceride content, as opposed to stained hepatocyte vacuoles in a microscope section. Retention of polyethyleneglycol (PEG) within renal tubular epithelium may be better evaluated by direct measurement of PEG in tissue samples. Where morphometry of immunohistochemistry sections is being considered, tissue digestion followed by fluorescence-activated cell sorting (FACS) could be considered as an alternative, since this counts most or all of the cells in a tissue sample, as opposed to microscope sections which sample a very small fraction of the total, and with no known relationship between the 2D cell counts and their numbers in the organ as a whole. Sometimes, very simple, standard end points will suffice: cardiac weight as an alternative to morphometry of cardiomyocytes, or ventricular wall thickness in cases of cardiac hypertrophy.

8.4.6   Limitations of Pathology Related to Subjectivity and Pathologist Error

The importance of subjectivity in toxicologic pathology is magnified by the fact that not only the diagnostic data, but the anatomic pathologist's interpretation of them in their narrative report represent raw data within a study (United States Federal Register, 1987). This means that once signed, the anatomic pathology report is not readily open to reinterpretation, so it is critical that the sponsor toxicologist review and provide input at the draft stage. The difficulty for toxicologists who do not have in-house pathology support is that anatomic-pathology morphological diagnosis is not something that non-specialists can generally participate in meaningfully, and the subsequent interpretation of these data can be extremely challenging for those without formal pathology training and direct experience in laboratory-animal pathology. Clinical pathology data interpretation is similarly something that is difficult for non-veterinarians without specialist training and experience. Because these data are something of a 'black box', non-pathologists can sometimes feel ill-equipped to critically review or challenge them – or the interpretation with which they are presented. However, an understanding of the common sources of potential error in diagnosis and interpretation can assist the toxicologist in recognising potential red flags related to pathology data generation and interpretation, and in managing any issues early in the study reporting process. It is also useful for toxicologists to be aware of the limitations of pathology data relating to subjectivity, in order to avoid unnecessary risk when considering adding pathology end points to nonstandard toxicology studies.

8.4.7   Anatomic Pathology Error/Missed Findings

Much of medical science must rely on the subjective judgement of individual practitioners. Key areas include radiology, diagnostic pathology and, of course, clinical diagnosis generally. Whilst everyone recognises that experts may often disagree on a particular difficult diagnosis, it is less well appreciated how often even relatively straightforward diagnoses are missed or misdiagnosed in human clinical practice (e.g. Robinson, 1997; Berner and Graber, 2008; Schiff et al., 2009). The degree to which errors occur in toxicologic pathology has not been quantified, but data exist for other perceptual specialties such as radiology and medical diagnostic pathology that require a combination of careful observation and subjective judgement.

Broad reviews of error in radiology have demonstrated a range of 2–20% of radiology cases with 'clinically significant or major error' (Goddard et al., 2001). A study of the 'miss rate' of non-small-cell lung cancer within a complete set of chest radiographs found a missed diagnosis in 49/396 patients (19%) (Quekel et al., 1999). This was considered a low error rate by the authors. Another study of 182 radiographic cases found 126 incidences of perceptual errors (68%): false positives, false negatives and misclassifications (Renfrew et al., 1992). There have been many studies of error rates in mammography: miss rates for breast carcinoma vary from 10 to over 40% (e.g. Majid et al., 2003; Harvey et al., 1993). In medical pathology, expert review of diagnostic samples can demonstrate substantial discordance: a recent study found clinically meaningful diagnostic discordance in 38/407 breast biopsy cases (9.6%) (Feng et al., 2014). Certain diagnoses, such as lymphomas and sarcomas, are subject to much higher discordance rates (Kim et al., 1982; Harris et al., 1991).

Although no formal studies have been published, it is likely that toxicologic pathology shares similar error/discordance rates, and anecdotal evidence from toxicologic pathologists themselves supports this. An understanding of this possibility by the toxicologist – and of how to address it – matters. A missed or mistaken diagnosis in human medicine typically only has repercussions for the individual patient affected. In contrast, missed or mistaken diagnoses in toxicologic pathology can have repercussions affecting multiple individuals – potentially many millions of individuals, who may be harmed by the toxicological effects of a drug or chemical, or alternatively, denied a potentially effective therapy based on spurious histopathology findings that confound drug development.

Further errors in pathology findings can occur as a result of a pathologist's unfamiliarity with a lesion, data input and reporting problems in computer pathology reporting systems, the use of different criteria for tumour classification and the use of different terminologies for the same lesion. Some lesions have several different synonymous names, all of which describe the same finding (e.g. 'sinus dilatation' in the lymph node is also called 'cystic degeneration', 'cystic ectasis', 'lymphangiectasis' and 'lymphatic cysts'; Figure 8.1).

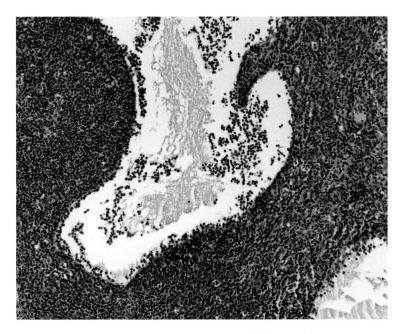

Figure 8.1 Sinus dilatation in the mouse lymph node – also called 'cystic degeneration', 'cystic ectasis', 'lymphangiectasis' and 'lymphatic cysts'.

8.4.8 Subjectivity and Pathologist Variability

More subtle limitations of pathology related to subjectivity come from differences between pathologists in grading lesion severity and/or recording tissue changes that they consider to fall 'within normal limits'. The choice of the wrong thresholds for diagnosing lesion severity at a particular grade, or for diagnosing it at all, can mask a test-article effect, or suggest a test-article effect where this is none (McInnes and Scudamore, 2014).

Similar difficulties can arise associated with 'lumping' versus 'splitting' by the study pathologist. Lumping refers to the combination of a number of different morphological changes into a single diagnosis (see Chapter 2), normally with the goal of capturing the full spectrum of lesions associated with a single pathogenesis. Splitting, on the other hand, is the specification of several related changes as individual diagnoses, normally with the goal of better characterising the changes directly in the study data (as opposed to an explanatory narrative in the pathology report).

An example might be 'focal inflammation' in the prostate gland. Such a diagnosis could refer simply to focal accumulation of inflammatory leukocytes at one or more sites in the prostate gland, but as a 'lumped' diagnosis it can also refer to a spectrum of changes where inflammation is perceived to be the primary pathogenic mechanism, but additional features (depending on the severity or chronicity) could include focal oedema, haemorrhage, epithelial necrosis, epithelial hyperplasia, fibrosis and/or atrophy. The advantage of lumping is that the incidence of the lumped finding reflects the true number of animals affected by what is really a single test-article effect, regardless of severity or chronicity, so the incidence can be directly compared across animals sacrificed early or late in the study, or even at the end of the recovery period. One disadvantage is that this limits any distinctions between adverse and nonadverse forms of the same process– for instance, uncomplicated inflammation might be considered nonadverse, whereas inflammation accompanied by other changes might be considered adverse, and review of the lesion incidence table does not provide any data to support this distinction. Uncomplicated inflammation may also be present as a spontaneous change in controls, such that the magnitude of the test-article effect is somewhat or even completely masked when considering the relative incidence across groups. Other problems can arise if the description in the narrative report is inadequate, such that the true extent of the secondary changes arising from the 'inflammation' is not clear to the reader and results in an inappropriate human risk assessment.

Splitting the test-article effect into several diagnoses directly reflects the morphological changes present on the slide, so that there is no doubt as to what tissue changes are taking place as a result of the test article. It can also be argued that it is more consistent with GLP data recording, since lumping implies some level of interpretation that cannot be readily reconstructed from the data and which should potentially be open to challenge. That said, splitting may ultimately be less informative for the reader, particularly if the relationship of the multiple test-article effects is not adequately communicated by the pathologist in his or her narrative report. Some low-incidence features may appear to lack a dose response, and the tables can give a false impression of recovery, since a more chronic form of the original lesion will be diagnosed differently so that it may appear to have recovered when in reality it has become more advanced.

8.5 Managing Risk Associated with Subjectivity and the Potential for Pathologist Error

8.5.1 Choice of Study Pathologist

Whilst there are certainly some excellent toxicologic pathologists with no clinical veterinary training or any of the specific anatomic/clinical pathology qualifications that certify diagnostic competency, there is no simple way for a toxicologist or anyone else commissioning a study to identify such individuals among others who may not be suitably qualified or experienced to interpret anatomic or clinical pathology data in a toxicology study. Requesting board certification (or the local equivalent) is helpful, since it ensures a minimum standard of competency and is recommended for toxicologists placing studies externally, without internal pathology support.

However, toxicologists should also bear in mind that most of the training programmes and examination content for anatomic or clinical pathology board certification are based largely on diagnostic pathology of natural disease. Whilst this is an essential baseline from which to understand and recognise pathological processes, further experience under active mentorship is required to become reliably competent in toxicologic-pathology diagnosis and interpretation. Specifying a minimum number of years of regulatory study experience (regulatory studies include comprehensive tissue lists) in addition to board certification is therefore also advisable.

Conversely, senior pathologists who are managing other pathologists and not routinely reading studies may also be lacking sufficient hands-on experience and are not necessarily the best choice. Pathologists who do principally investigative or efficacy work that does not regularly involve comprehensive tissue lists, subtle lesions or ageing animals may also be less suitable for primary reads or peer review of full-tissue regulatory studies. Where a clinical pathologist is not available to interpret the clinical pathology, an anatomic pathologist is acceptable, since in both the United States and Europe, anatomic-pathology board certification includes a clinical pathology component (ACVP, 2015; ECVP, 2015).

8.5.2 Peer Review

Risks associated with the generation and interpretation of subjective data should be always be managed by the use of peer review, where a second pathologist reviews the work of the primary pathologist on a study. A typical peer review is guided by a specific SOP, and at a minimum consists of the re-evaluation of a proportion of the slides from the high-dose animals, all test-article-related findings and review of the narrative report in the context of the other study data. In a typical regulatory toxicology study, other study data usually include organ weights, clinical pathology, clinical observations, ophthalmology and, for large animals, electrocardiogram reports. Where possible, a peer review is usually best conducted by a pathologist of greater experience than the study pathologist; this can be a sponsor pathologist, another pathologist from the same laboratory or a third party selected by the sponsoring toxicologist.

Surprisingly, there is no regulatory requirement for peer review of toxicologic pathology data. An informal assessment by an FDA conference speaker in 2012 suggested that approximately half of toxicology studies received by them have been peer-reviewed, and that peer review is generally desirable. Recent marketing approval documents published by the FDA frequently mention whether pivotal toxicology studies have received pathology peer review. Whilst peer review does not guarantee that mistakes will not occur in pathology diagnosis, it is generally accepted that it significantly improves the quality of the data obtained (Morton et al., 2010).

8.5.3 Review of the Anatomic Pathology Data

There are certain situations in which review of the data can signal to the toxicologist (and other reviewers) that the anatomic pathology data (the actual diagnoses listed in the study tables, not the subsequent interpretation) he or she is receiving may require further expert review. This is not a comprehensive list, but some of the most common situations where review is advisable include:

- diagnoses consistent with common artefacts;
- large differences in the incidence of findings between dose groups, irrespective of dose;
- degenerative changes in the male reproductive tract when age and testicular weights suggest peripubertal status;
- absence of expected test article-related changes from the tables;
- lack of microscopic correlates for gross necropsy findings and lack of clinical pathology correlates for microscopic findings (and vice versa).

Where any of these situations is suggested by the data, the toxicologist should discuss the reasons for it with the study pathologist and/or the peer reviewer. The tissues most prone to diagnostic error by inexperienced pathologists include the reproductive tract, bone, eye and brain.

8.5.4 Review of Anatomic Pathology Data Interpretation

As mentioned earlier, in regulatory toxicology studies, the interpretation of the anatomic pathology data by the pathologist once the slide-evaluation phase is complete has a special status: the signed pathologist narrative is considered raw data. This means that the limitations imposed by the ability and experience of the pathologist (and peer reviewer) can irrevocably affect the outcome of the study. Many of the risk-management strategies suggested in this section to ensure the best possible anatomic pathology diagnostic data are applicable to the management of risk associated with pathology data interpretation. However, the toxicologist is in a position to play a larger role once the initial diagnostic data have been generated, and there is potentially much to be gained by conducting a thoughtful review of the pathology report and discussing any potential interpretation shortcomings with the pathologist and peer reviewer prior to finalization of the contributor report.

One of the most common situations in which data are overinterpreted occurs when the pathology change designated 'test article-related' is one that represents an increase in severity or incidence over that in the control group and/or is known to occur reasonably commonly as a spontaneous change. Typical examples include small inflammatory infiltrates in the liver in rodents and focal renal tubular degeneration/regeneration in most species (see Chapter 3), where there is plenty of scope for random differences in incidence/severity between groups to be disproportionately distributed more in the treated groups and thus to appear test article-related. This can be a particular problem in large-animal studies where the number of animals is small. Less experienced patholo-gists sometimes limit themselves to comparing treated animals with the concurrent con-trols in the study, and if they can distinguish the control from the treated animals (including when evaluating slides in a blinded fashion), they conclude that any difference must represent an effect of the test article. More experienced pathologists tend to better appreciate the full range of spontaneous change and what is truly 'within normal limits', regardless of the relative incidence of common findings amongst control and treated groups. That said, as discussed earlier, even with flawless anatomic pathology diagnosis and interpretation, common spontaneous and background changes can potentially con-found the interpretation of general toxicology studies, which have no inbuilt allowance for multiple tested end points and often a lack of reliable historical control data.

Another situation that can signal overinterpretation of anatomic or clinical pathology data is where the narrative describes findings that are consistent with nonspecific signs of stress/moribundity/reduced food consumption as primary test article-related effects. These can include certain patterns of organ-weight changes, thymic atrophy, atrophic changes in reproductive organs and a variety of clinical pathology changes, such as pre-renal increases in blood urea nitrogen (sometimes also creatinine and phosphorus), reductions in lymphocytes or reticulocytes and either increases or decreases in serum glucose concentrations. If other study data support an indirect aetiology, it is useful for the toxicologist to discuss the interpretation with the anatomic or clinical pathologist in order to understand the grounds for designating the various test article-related changes as 'direct' or 'indirect'.

By contrast, underinterpretation occurs when data that signal a test-article effect are mistakenly interpreted as background noise by the anatomic or clinical pathologist. Circumstances in which this is more likely to occur include situations where the pathol-ogist and/or peer reviewer are unaware of the pharmacological mechanism of action and/or have not adequately researched potential toxicology risks arising from that mechanism of action. It can also occur with inappropriate 'thresholding', such that too wide a range of severity is encompassed within a single severity grade (or is designated 'within normal limits') so that true test article-related differences are not discernible from the incidence tables (McInnes and Scudamore, 2014).

As with questions about the actual pathology diagnoses, where a toxicologist is con-cerned about pathology interpretation, detailed discussions should be conducted with both the pathologist and the peer reviewer to explore the possibility of alternative inter-pretations, in order to ensure that all viewpoints have been considered and to under-stand the reasons (if not articulated in the pathology report) for the interpretation that was given by the pathologist.

The subjectivity risk, particularly for unfamiliar end points and/or in unfamiliar mod-els, means that anatomic and clinical pathology evaluation may be best avoided alto-gether where alternatives exist. Where not required to fulfil the objectives of the study, toxicologists (or others designing a study) should consider carefully before adding on pathology end points as nice-to-have or supportive data on nonstandard studies. They must determine clearly the objective of the study and collect only data directly pertain-ing to that goal. If the applicability of pathology end points cannot be determined before other data are collected (such as need for elucidating cause of death/moribundity), save tissues/samples, with examination made contingent on further study data, rather than evaluating them prospectively.

References

ACVP (2015) Scope, Knowledge and Requirements for the ACVP Certifying Examinations in Anatomic Pathology and Clinical Pathology. American College of Veterinary Pathologists. Available from: http://www.acvp.org/index.php/en/2014-11-07-22-05-00/scope-knowledge-and-requirements-for-the-acvp-certifying-examinations-in-anatomic-pathology-and-clinical-pathology (last accessed July 29, 2016).

Adams, E.T. and Crabbs, T.A. (2013) Basic approaches in anatomic toxicologic pathology. In: Haschek, W.M., Rousseaux, C.G. and Walig, M.A. (eds). Haschek and Rousseaux's Handbook of Toxicologic Pathology, 3rd edn, Academic Press, Cambridge, MA, pp. 153–65.

Babaya, K., Takahashi, S., Momose, H., Matsuki, H., Sasaki, K., Samma, S., Ozono, S., Hirao, Y. and Okajima, E. (1987) Effects of single chemotherapeutic agents on development of urinary bladder tumor induced by N-butyl-N-(4-hydroxybutyl) nitrosamine (BBN) in rats. *Urological Research*, 15(6), 329–34.

Bendele, A.M. and White, S.L. (1987) Early histopathologic and ultrastructural alterations in femorotibial joints of partial medial meniscectomized guinea pigs. *Veterinary Pathology*, 24(5), 436–43.

Bendele, A., McComb, J., Gould, T., McAbee, T., Senneloo, G., Chlipala, E. and Guy, M. (1999) Animal models of arthritis: relevance to human disease. *Toxicologic Pathology*, 27(1), 134–42.

Berner, E.S. and Graber, M.L. (2008) Overconfidence as a cause of diagnostic error in medicine. *American Journal of Medicine*, 121(5), S2–23.

Boyce, J.T., Boyce, R.W. and Gundersen, H.J. (2010a) Choice of morphometric methods and consequences in the regulatory environment. *Toxicologic Pathology*, 38(7), 1128–33.

Boyce, R.W., Dorph-Petersen, K.A., Lyck, L. and Gundersen, H.J.G. (2010b) Design-based stereology introduction to basic concepts and practical approaches for estimation of cell number. *Toxicologic Pathology*, 38(7), 1011–25.

Calcutt, N.A., Cooper, M.E., Kern, T.S. and Schmidt, A.M. (2009) Therapies for hyperglycaemia-induced diabetic complications: from animal models to clinical trials. *Nature Reviews Drug Discovery*, 8(5), 417–30.

Cannon, C.P. and Cannon, P.J. (2012) COX-2 inhibitors and cardiovascular risk. *Science*, 336(6087), 1386–7.

Chamanza, R., Marxfeld, H.A., Blanco, A.I., Naylor, S.W. and Bradley, A.E. (2010) Incidences and range of spontaneous findings in control cynomolgus monkeys (Macaca fascicularis) used in toxicity studies. *Toxicologic Pathology*, 38(4), 642–57.

Chevalier, R.L., Forbes, M.S. and Thornhill, B.A. (2009) Ureteral obstruction as a model of renal interstitial fibrosis and obstructive nephropathy. *Kidney International*, 75(11), 1145–52.

Christensen, R., Kristensen, P.K., Bartels, E.M., Bliddal, H. and Astrup, A. (2007) Efficacy and safety of the weight-loss drug rimonabant: a meta-analysis of randomised trials. *Lancet*, 370(9600), 1706–13.

Creasy, D.M. (2003) Evaluation of testicular toxicology: a synopsis and discussion of the recommendations proposed by the Society of Toxicologic Pathology. *Birth Defects Research, Part B: Developmental and Reproductive Toxicology*, 68(5), 408–15.

Daly, A.K. and Day, C.P. (2012) Genetic association studies in drug-induced liver injury. *Drug Metabolism Reviews*, 44(1), 116–26.

Eastwood, D., Findlay, L., Poole, S., Bird, C., Wadhwa, M., Moore, M., Burns, C., Thorpe, R. and Stebbings, R. (2010) Monoclonal antibody TGN1412 trial failure explained by species differences in CD28 expression on CD4+ effector memory T-cells. *British Journal of Pharmacology*, 161(3), 512–26.

ECVP. (2015) Exam format FAQ. European College of Veterinary Pathologists. Available from: http://www.ecvpath.org/exam-format-faq/ (last accessed July 29, 2016).

FDA. (2000) Center for Drug Evaluation and Research Application Number: 21-081. Pharmacology Review(s). Available from: http://www.accessdata.fda.gov/drugsatfda_ docs/nda/2000/21081_Lantus_pharmr_P1.pdf (last accessed July 29, 2016).

FDA. (2001) Guidance for Industry. Statistical Aspects of the Design, Analysis and Interpretation of Chronic Rodent Carcinogenicity Studies of Pharmaceuticals. US Food and Drug Administration, Rockville, MD.

FDA. (2009) Guidance for Industry. Drug-Induced Liver Injury: Premarketing Clinical Evaluation. US Food and Drug Administration, Rockville, MD.

FDA. (2013) FDA Briefing Document NDA 202293. US Food and Drug Administration, Rockville, MD. Available from: http://www.fda.gov/downloads/AdvisoryCommittees/ CommitteesMeetingMaterials/Drugs/ EndocrinologicandMetabolicDrugsAdvisoryCommittee/UCM378076.pdf (last accessed July 29, 2016).

FDA. (2014) Advancing regulatory science. 1. Modernize toxicology to enhance product safety. US Food and Drug Administration, Rockville, MD. Available from: http://www. fda.gov/scienceresearch/specialtopics/regulatoryscience/ucm268111.htm (last accessed July 29, 2016).

Feng, S., Weaver, D.L., Carney, P.A., Reisch, L.M., Geller, B.M., Goodwin, A., Rendi, M.H., Onega, T., Allison, K.H., Tosteson, A.N., Nelson, H.D., Longton, G., Pepe, M. and Elmore, J.G. (2014) A framework for evaluating diagnostic discordance in pathology discovered during research studies. *Archives of Pathology & Laboratory Medicine*, 138(7), 955–61.

Fukushima, S., Kurata, Y., Shibata, M.A., Ikawa, E. and Ito, N. (1984) Promotion by ascorbic acid, sodium erythorbate and ethoxyquin of neoplastic lesions in rats initiated with N-butyl-N-(4-hydroxybutyl) nitrosamine. *Cancer Letters*, 23(1), 29–37.

Goddard, P., Leslie, A., Jones, A., Wakeley, C. and Kabala, J. (2001) Error in radiology. *British Journal of Radiology*, 74(886), 949–51.

Goedken, M.J., Kerlin, R.L. and Morton, D. (2008) Spontaneous and age-related testicular findings in beagle dogs. *Toxicologic Pathology*, 36(3), 465–71.

Graham, D.J. (2006) COX-2 inhibitors, other NSAIDs, and cardiovascular risk: the seduction of common sense. *JAMA*, 296(13), 1653–6.

Greaves, P. (2012) Nervous system and special sense organs. In: Greaves, P. (ed.) Histopathology of Preclinical Toxicity Studies, 4th edn, Elsevier, Amsterdam, p. 801.

Greaves, P., Williams, A. and Eve, M. (2004) First dose of potential new medicines to humans: how animals help. *Nature Reviews Drug Discovery*, 3(3), 226–36.

Gundersen, H.J.G., Mirabile, R., Brown, D. and Waite Boyce, R. (2013) Stereological principles and sampling techniques for toxicologic pathologists. In: Haschek, W.M., Rousseaux, C.G. and Walig, M.A. (eds). Haschek and Rousseaux's Handbook of Toxicologic Pathology, 3rd edn, Academic Press, Cambridge, MA, p. 216.

Gunson, D., Gropp, K.E. and Varela, A. (2013) Bone and joints. In: Haschek, W.M., Rousseaux, C.G. and Walig, M.A. (eds). Haschek and Rousseaux's Handbook of Toxicologic Pathology, 3rd edn, Academic Press, Cambridge, MA, p. 2761.

Hardisty, J.F. (1985) Factors influencing laboratory animal spontaneous tumor profiles. *Toxicologic Pathology*, 13(2), 95–104.

Harris, M., Hartley, A.L., Blair, V., Birch, J.M., Banerjee, S.S., Freemont, A.J., McClure, J. and McWilliam, L.J. (1991) Sarcomas in North West England: I. Histopathological peer review. *British Journal of Cancer*, 64(2), 315.

Harvey, J.A., Fajardo, L.L. and Innis, C.A. (1993) Previous mammograms in patients with impalpable breast carcinoma: retrospective vs blinded interpretation. 1993 ARRS President's Award. *American Journal of Roentgenology*, 161(6), 1167–72.

Hawkey, C.J. (1999) COX-2 inhibitors. *Lancet*, 353(9149), 307–14.

Horvath, C., Andrews, L., Baumann, A., Black, L., Blanset, D., Cavagnaro, J., Hasting, K.L., Hutto, D.L., MacLachlan, T.K., Milton, M., Reynolds, T., Roberts, S., Rogge, M., Sims, J., Treacy, G., Warner, G. and Green, J.D. (2012) Storm forecasting: additional lessons from the CD28 superagonist TGN1412 trial. *Nature Reviews. Immunology*, 12(10), 740.

ICH. (2001) Guidance for Industry. S7A Safety Pharmacology Studies for Human Pharmaceuticals. US Food and Drug Administration, Rockville, MD.

ICH. (2006) Guidance for Industry. S8 Immunotoxicity Studies for Human Pharmaceuticals. US Food and Drug Administration, Rockville, MD.

Ince, T.A., Ward, J.M., Valli, V.E., Sgroi, D., Nikitin, A.Y., Loda, M., Griffey, S.M., Crum, C.P., Crawford, J.M., Bronson, R.T. and Cardiff, R.D. (2008) Do-it-yourself (DIY) pathology. *Nature Biotechnology*, 26(9), 978–9.

Keiser, N.W. and Engelhardt, J.F. (2011) New animal models of cystic fibrosis: what are they teaching us? *Current Opinion in Pulmonary Medicine*, 17(6), 478–83.

Kim, H., Zelman, R.J., Fox, M.A., Bennett, J.M., Berard, C.W., Butler, J.J., Byrne, G.E., Dorfman, R.F., Hartsock, R.J., Lukes, R.J. and Mann, R.B. (1982) Pathology panel for lymphoma clinical studies: A comprehensive analysis of cases accumulated since its inception. *Journal of the National Cancer Institute*, 68(1), 43–67.

King, A.J. (2012) The use of animal models in diabetes research. *British Journal of Pharmacology*, 166(3), 877–94.

Ku, W.W., Pagliusi, F., Foley, G., Roesler, A. and Zimmerman, T. (2010) A simple orchidometric method for the preliminary assessment of maturity status in male cynomolgus monkeys (Macaca fascicularis) used for nonclinical safety studies. *Journal of Pharmacological and Toxicological Methods*, 61(1), 32–7.

Kumar, V., Abbas, A.K. and Aster, J.C. (2015) Hemodynamic disorders, thrombosis and shock. In: Kumar, V., Abbas, A.K. and Aster, J.C. (eds). Robbins & Cotran Pathologic Basis of Disease, Elsevier, Philadelphia, PA, pp. 131–4.

Lanning, L.L., Creasy, D.M., Chapin, R.E., Mann, P.C., Barlow, N.J., Regan, K.S. and Goodman, D.G. (2002) Recommended approaches for the evaluation of testicular and epididymal toxicity. *Toxicologic Pathology*, 30(4), 507–20.

Like, A.A. and Rossini, A.A. (1976) Streptozotocin-induced pancreatic insulitis: new model of diabetes mellitus. *Science*, 193(4251), 415–17.

Lowenstine, L.J. (2003) A primer of primate pathology: lesions and nonlesions. *Toxicologic Pathology*, 31(1 Suppl.), 92–102.

Lubet, R.A., Fischer, S.M., Steele, V.E., Juliana, M.M., Desmond, R. and Grubbs, C.J. (2008) Rosiglitazone, a PPAR gamma agonist: potent promoter of hydroxybutyl (butyl) nitrosamine-induced urinary bladder cancers. *International Journal of Cancer*, 123(10), 2254–9.

Majid, A.S., de Paredes, E.S., Doherty, R.D., Sharma, N.R. and Salvador, X. (2003) Missed breast carcinoma: pitfalls and pearls. *Radiographics*, 23(4), 881–95.

McInnes, E.F. (2012) Background Lesions in Laboratory Animals: A Color Atlas, Elsevier, Edinburgh.

McInnes, E.F. and Scudamore, C.L. (2014) Review of approaches to the recording of background lesions in toxicologic pathology studies in rats. *Toxicology Letters*, 229(1), 134–43.

Mendis-Handagama, S.M.L.C. and Ewing, L.L. (1990) Sources of error in the estimation of Leydig cell numbers in control and atrophied mammalian testes. *Journal of Microscopy*, 159(1), 73–82.

Morgan, S.J., Elangbam, C.S., Berens, S., Janovitz, E., Vitsky, A., Zabka, T. and Conour, L. (2013) Use of animal models of human disease for nonclinical safety assessment of novel pharmaceuticals. *Toxicologic Pathology*, 41(3), 508–18.

Morton, D., Sellers, R. S., Barale-Thomas, E., Bolon, B., George, C., Hardisty, J.F., Irizarry, A., McKay, J.S., Odin, M. and Teranishi, M. (2010) Recommendations for pathology peer review. *Toxicologic Pathology*, 38(7), 1118–27.

Okada, M., Sano, F., Ikeda, I., Sugimoto, J., Takagi, S., Sakai, H. and Yanai, T. (2009) Fenofibrate-induced muscular toxicity is associated with a metabolic shift limited to type-1 muscles in rats. *Toxicologic Pathology*, 37(4), 517–20.

Olar, T.T., Amann, R.P. and Pickett, B.W. (1983) Relationships among testicular size, daily production and output of spermatozoa, and extragonadal spermatozoal reserves of the dog. *Biology of Reproduction*, 29(5), 1114–20.

Olson, H., Betton, G., Robinson, D., Thomas, K., Monro, A., Kolaja, G., Lilly, P., Sanders, J., Sipes, G., Bracken, W., Dorato, M., Van Deun, K., Smith, P., Berger, B. and Heller, A. (2000) Concordance of the toxicity of pharmaceuticals in humans and in animals. *Regulatory Toxicology and Pharmacology*, 32(1), 56–67.

Perry, R., Farris, G., Bienvenu, J.G., Dean C. Jr, Foley, G., Mahrt, C. and Short, B.; Society of Toxicologic Pathology. (2013) Society of toxicologic pathology position paper on best practices on recovery studies the role of the anatomic pathologist. *Toxicologic Pathology*, 41(8), 1159–69.

Pettersen, J.C., Litchfield, J., Neef, N., Schmidt, S.P., Shirai, N., Walters, K.M., Enerson, B.E., Chatman, L.A. and Pfefferkorn, J.A. (2014) The relationship of glucokinase activator-induced hypoglycemia with arteriopathy, neuronal necrosis, and peripheral neuropathy in nonclinical studies. *Toxicologic Pathology*, 42(4), 696–708.

Pfeifer, M., Boncristiano, S., Bondolfi, L., Stalder, A., Deller, T., Staufenbiel, M., Mathews, P.M. and Jucker, M. (2002) Cerebral hemorrhage after passive anti-Aβ immunotherapy. *Science*, 298(5597), 1379.

Quekel, L.G., Kessels, A.G., Goei, R. and van Engelshoven, J.M. (1999) Miss rate of lung cancer on the chest radiograph in clinical practice. *Chest*, 115(3), 720–4.

Renfrew, D.L., Franken, E.A. Jr, Berbaum, K.S., Weigelt, F.H. and Abu-Yousef, M.M. (1992) Error in radiology: classification and lessons in 182 cases presented at a problem case conference. *Radiology*, 183(1), 145–50.

Richardson, F.C., Boucheron, J.A., Dyroff, M.C., Popp, J.A. and Swenberg, J.A. (1986) Biochemical and morphologic studies of heterogeneous lobe responses in hepatocarcinogenesis. *Carcinogenesis*, 7(2), 247–51.

Robinson, P.J. (1997) Radiology's Achilles' heel: error and variation in the interpretation of the Röntgen image. *British Journal of Radiology*, 70(839), 1085–98.

Schiff, G.D., Hasan, O., Kim, S., Abrams, R., Cosby, K., Lambert, B.L., Elstein, A.S., Hasler, S., Kabongo, M.L., Krosnjar, N., Odwazny, R., Wisniewski, M.F. and McNutt, R.A.

(2009) Diagnostic error in medicine: analysis of 583 physician-reported errors. *Archives of Internal Medicine*, 169(20), 1881–7.

Smedley, J.V., Bailey, S.A., Perry, R.W. and O'Rourke, C.M. (2002) Methods for predicting sexual maturity in male cynomolgus macaques on the basis of age, body weight, and histologic evaluation of the testes. *Journal of the American Association for Laboratory Animal Science*, 41(5), 18–20.

Suntharalingam, G., Perry, M.R., Ward, S., Brett, S.J., Castello-Cortes, A., Brunner, M.D. and Panoskaltsis, N. (2006) Cytokine storm in a phase 1 trial of the anti-CD28 monoclonal antibody TGN1412. *New England Journal of Medicine*, 355(10), 1018–28.

Tirmenstein, M., Horvath, J., Graziano, M., Mangipudy, R., Dorr, T., Colman, K., Zinker, B., Kirby, M., Cheng, P.T., Patrone, L., Kozlosky, J., Reilly, T.P., Wang, V. and Janovitz, E. (2015) Utilization of the Zucker diabetic fatty (ZDF) rat model for investigating hypoglycemia-related toxicities. *Toxicologic Pathology*, 43(6), 825–37.

United States Federal Register. (1987) *Preamble to the Good Laboratory Practice Regulations 1987*, 52(172), 33 768–82.

Wanibuchi, H., Yamamoto, S., Chen, H., Yoshida, K., Endo, G., Hori, T. and Fukushima, S. (1996) Promoting effects of dimethylarsinic acid on N-butyl-N-(4-hydroxybutyl) nitrosamine-induced urinary bladder carcinogenesis in rats. *Carcinogenesis*, 17(11), 2435–9.

Westwood, F.R., Bigley, A., Randall, K., Marsden, A.M. and Scott, R.C. (2005) Statin-induced muscle necrosis in the rat: distribution, development, and fibre selectivity. *Toxicologic Pathology*, 33(2), 246–57.

Yu, Y., Ricciotti, E., Scalia, R., Tang, S.Y., Grant, G., Yu, Z., Landesberg, G., Crichton, I., Wu, W., Puré, E., Funk, C.D. and FitzGerald, G.A. (2012) Vascular COX-2 modulates blood pressure and thrombosis in mice. *Science Translational Medicine*, 4(132), 132ra54.

Glossary

Abscess is a circumscribed collection of pus within a fibrous tissue capsule.

Adenocarcinoma is a malignant tumour made up of epithelial glandular cells.

Aetiology is the cause of a disease or abnormal clinical signs.

Alopecia is the loss of hair.

Amyloid is a dense protein which can accumulate in tissues during some chronic infectious, inflammatory, immune or neoplastic diseases.

Anaemia is low haemoglobin or erythrocyte levels in the blood.

Anaphylaxis is a sudden type 1 hypersensitivity reaction to a stimulus.

Anaplasia is a reversion to a more primitive embryonic cell type seen in malignant tumours.

Anasarca is the accumulation of oedematous fluid in the subcutaneous tissues.

Antigen is a foreign molecule (generally) able to stimulate the production of specific antibodies.

Apoptosis or **programmed cell death** is a regulated cell suicide programme.

Ascites is the accumulation of oedematous fluid within the peritoneal cavity.

Atrophy is a process where a cell or organ reduces its mass and size.

Autolysis is the enzymatic digestion of cells, particularly after death.

Basophil is a white blood cell with cytoplasmic granules that stain blue in histological sections.

Benign tumour is a localised neoplasm that generally does not spread (metastasis).

Carcinogenesis is the process of tumour development.

Carcinoma is a malignant tumour made up of epithelial cells.

Cataract is the degeneration of the lens of the eye.

Chemotaxis is the process whereby inflammatory cells move into an area of inflammation.

Cholestasis is the cessation of the flow of bile, which may result in jaundice.

Chronic progressive nephropathy (CPN) is a common spontaneous kidney disease of ageing rats and mice.

Cluster of differentiation (CD) is a marker of lymphocytes and other cells. Leucocytes are differentiated by their cell surface molecules, which are identified by monoclonal antibodies.

Coagulation is the rapid production of a local plug at the site of blood vessel injury.

Coagulative necrosis is a form of necrosis in which the dead tissues form an amorphous mass.

Pathology for Toxicologists: Principles and Practices of Laboratory Animal Pathology for Study Personnel,
First Edition. Edited by Elizabeth McInnes.
© 2017 John Wiley & Sons Ltd. Published 2017 by John Wiley & Sons Ltd.

Coagulopathy is a disease affecting the clotting system that generally causes widespread haemorrhage.

Complement is a system consisting of proteins that mediate a series of reactions (increased vascular permeability, chemotaxis, opsonisation) that defend against microbial agents.

Conjunctivitis is the inflammation of the skin around the eye.

Cytology is the examination of cells in body fluids or from fine needle aspirates under a light microscope. The cells are not processed into wax blocks.

Diagnostic drift occurs in long-term studies where the pathologist may start to over- or underrecord certain background lesions with respect to their starting point over time.

Disseminated intravascular coagulation (DIC) is a condition where widespread endothelial damage leads to multiple small thromboses and a predisposition to haemorrhage.

Dry gangrene is the death of tissue due to restriction of the blood supply, which often affects extremities such as the tips of the ears.

Dysplasia is abnormal tissue development.

Ectopic means out of place and is generally a piece of tissue in an abnormal position that is present from birth.

Embolus is a substance (e.g. cancer cells or air) that is able to move through the bloodstream and lodge in a blood vessel and cause an obstruction.

Eosinophils are motile white blood cells with segmented nuclei and pink granules.

Erythema is reddening of the skin.

Exophthalmos is a bulging and protruding eye.

Extramedullary haemopoiesis (EMH) is the presence of immature, nucleated red blood cells, often in response to a regenerative anaemia.

Exudate is an accumulation of a protein-rich fluid due to porous blood vessels.

Fibrin is derived from fibrinogen and is used to form blood clots.

Fibrosis is a repair process causing the thickening and scarring of tissue.

Granuloma is a mass with a centre made up of foreign material or microorganisms surrounded by large multinucleate macrophages, neutrophils and a layer of fibrosis.

Haematocrit is the portion of blood that is made up of cells.

Haematology is the study of cell types in the blood.

Haemolysis is the destruction of red blood cells.

Haemosiderin is a golden-brown pigment containing iron, which can be demonstrated in tissues using Perl's special stain.

Hyperpigmentation occurs when skin becomes darker.

Hyperplasia is an increase in the number of cells in a tissue.

Hypertrichosis is the excessive production of hair.

Hypertrophy is the increase in a cell or organ due to larger cell size rather than an increase in cell number.

Hypopigmentation occurs when the skin becomes lighter.

Hypoplasia is the decrease in the size of a tissue due to a reduction in cell number.

Hypotrichosis is the reduction in hair production.

Idiopathic is a lesion that does not have a definite cause.

Immunodeficiency occurs when an animal has an immune system that does not function properly.

Infarct is a localised area of necrosis caused by a lack of blood (i.e. oxygen) supply to an area of tissue.

Intussusception occurs when a portion of intestine telescopes into the lumen of the intestine adjacent to it.

Ischaemia is a state of reduced blood, that is, oxygen, in tissue.

Jaundice (or **icterus**) is the yellow discoloration of tissues and fluids because of an excess of bile pigments.

Keratinisation is the formation of the protein keratin, generally on an epithelial surface.

Keratitis is the inflammation of the cornea.

Leucocyte is a white blood cell.

Lumping is the process used by toxicological pathologists to capture a full spectrum of findings under a single lesion/finding.

Major histocompatibility complex (MHC) is a large group of genes including those encoding the class I and class II MHC molecules, which are involved in the presentation of antigen to T cells.

Malignant is a tumour that demonstrates invasion and metastasis.

Membrane attack complex (MAC) is a structure that traverses the target cell membrane and allows osmotic leakage from the cell.

Metaplasia is the abnormal transformation of a cell type from one type to another, for example, from olfactory epithelium to squamous epithelium.

Metastasis is the spread of cancer from its primary location to distant tissues or organs.

Mineralisation is the generally abnormal deposition of calcium in tissues.

Monocytes are circulating macrophages.

Necrosis occurs when a cell undergoes irreversible injury, and its causes include a reduced oxygen supply, oxygen-derived free radicals and physical agents.

Neoplasm is a cancer or new growth of cells.

Neutropaenia is a reduction of neutrophils in the blood.

Oedema is excess fluid in interstitial tissues or serous cavities.

Pancytopaenia is a decrease in all white and red blood cells and platelets in the blood.

Pathogenesis is the sequence of events as tissues respond to an aetiological agent.

Petechiae are small, pinpoint haemorrhages in a tissue.

Photosensitivity is the inflammation of the skin due to the action of ultraviolet light (e.g. sunlight) on photodynamic agents within the skin (generally a drug in laboratory animals).

Pneumonia is the inflammation of the lungs.

Purulent is any process that contains pus.

Pus is the collection of dead white blood cells, particularly neutrophils and necrotic tissue, and is often yellow or green in colour.

Sclerosis is the formation of large amounts of collagen and fibrous tissues.

Splitting is the process used by toxicological pathologists to separate different findings of a single lesion into multiple separate findings.

Thresholds are the arbitrary limits set by individual toxicological pathologists when deciding upon lesion severity below which no finding is recorded.

Thrombosis is the obstruction of a blood vessel due to the presence of a blood clot (thrombus).

Tumour is generally a neoplasm.

Ulcer is an area of damage due to full-thickness destruction of the overlying tissue such as the skin.

Vasculitis is the inflammation of blood vessels.

Index

Pathology for Toxicologists: Principles and Practices of Laboratory Animal Pathology for Study Personnel,
First Edition. Edited by Elizabeth McInnes.
© 2017 John Wiley & Sons Ltd. Published 2017 by John Wiley & Sons Ltd.

no observable adverse effect level
(NOAEL) 146–147, 149, 153–154
no observable effect level
(NOEL) 146–147
nothing abnormal detected (NAD) 26,
28–29
NSAID *see* nonsteroidal anti-inflammatory
drugs

o

oedema 46–48
Oil Red O 12
organ weights 136
ovarian cysts 103–104

p

packed cell volume (PCV) 119–120
pancreatitis 81–82
papillary necrosis 92–93
parathyroid gland hyperplasia 97,
100–101
PAS *see* periodic acid–Schiff
pathologist error 173, 176–178
pathology findings *see* recording pathology
data
pathology report 20
pathology techniques 1–22
animal considerations 2
artefacts 6
autolysis 4–5
biologicals 19–20
carcinogenicity 19
cassette label and blocking sheet 7–8,
11–12
concepts and definitions 1–2
confocal microscopy 16–17
continuous-infusion studies 18–19
decalcification 13
digital imaging 17
electron microscopy 15–16
euthanasia 4
fixation 5–6
glass slides 6–12
Good Laboratory Practice 4, 17–18, 20
image analysis 17
immunohistochemistry 13–14
inhalation studies 18

laser capture microscopy 16
lung inflation with fixative 5
macroscopic lesions 4
necropsy 2–5
pathology report 20
quality control 11–12
regulatory bodies 1, 17–18
in situ hybridisation 16
special histochemical stains 12–13
spermatocyte analysis 17
standard operating procedure 4, 18
study personnel 2
tissue crossreactivity 15
pathology working groups
(PWG) 32
PCR *see* polymerase chain reaction
PCV *see* packed cell volume
peer review 32, 176–177
periodic acid–Schiff (PAS) 12
peripheral nervous system
(PNS) 104–106
peritonitis 81
peroxisome proliferator-activated receptor
(PPAR) 52
peroxisome proliferator-activated receptor
(PPAR) agonists 13, 82, 87–88, 99
petechiae 46, 48, 75
phagocytosis 41–43
pharmacological effect 153
phaeochromocytoma 101–102
phospholipidosis
adversity 150–151
pathology techniques 15
target organ pathology 89
phosphorus 130
phosphotungstic acid haematoxylin
(PTAH) 12–13
photosensitivity 74
physiological adaptability 152
pituitary adenoma 100
platelets 124
pneumonia 88
PNS *see* peripheral nervous system
polymerase chain reaction (PCR) 16
polyps 81, 82
post mortem examinations *see* necropsy
potassium 130, 140